The Open University

MU120
Open Mathematics

GW00640938

Unit 13

Baker's dozen

MU120 course units were produced by the following team:

Gaynor Arrowsmith (Course Manager)

Mike Crampin (Author)

Margaret Crowe (Course Manager)

Fergus Daly (Academic Editor)

Judith Daniels (Reader)

Chris Dillon (Author)

Judy Ekins (Chair and Author)

John Fauvel (Academic Editor)

Barrie Galpin (Author and Academic Editor)

Alan Graham (Author and Academic Editor)

Linda Hodgkinson (Author)

Gillian Iossif (Author)

Joyce Johnson (Reader)

Eric Love (Academic Editor)

Kevin McConway (Author)

David Pimm (Author and Academic Editor)

Karen Rex (Author)

Other contributions to the text were made by a number of Open University staff and students and others acting as consultants, developmental testers, critical readers and writers of draft material. The course team are extremely grateful for their time and effort.

The course units were put into production by the following:

Course Materials Production Unit (Faculty of Mathematics and Computing)

Martin Brazier (Graphic Designer)	Diane Mole (Graphic Designer)
Hannah Brunt (Graphic Designer)	Kate Richenburg (Publishing Editor)
Alison Cadle (TEXOpS Manager)	John A.Taylor (Graphic Artist)
Jenny Chalmers (Publishing Editor)	Howie Twiner (Graphic Artist)
Sue Dobson (Graphic Artist)	Nazlin Vohra (Graphic Designer)
Roger Lowry (Publishing Editor)	Steve Rycroft (Publishing Editor)

This publication forms part of an Open University course. Details of this and other Open University courses can be obtained from the Student Registration and Enquiry Service, The Open University, PO Box 197, Milton Keynes MK7 6BJ, United Kingdom: tel. +44 (0)845 300 6090, email general-enquiries@open.ac.uk

Alternatively, you may visit the Open University website at http://www.open.ac.uk where you can learn more about the wide range of courses and packs offered at all levels by The Open University.

To purchase a selection of Open University course materials visit http://www.ouw.co.uk, or contact Open University Worldwide, Walton Hall, Milton Keynes MK7 6AA, United Kingdom, for a brochure: tel. +44 (0)1908 858793, fax +44 (0)1908 858787, email ouw-customer-services@open.ac.uk

The Open University, Walton Hall, Milton Keynes, MK7 6AA.

First published 1996. Second edition 2008.

Edited, designed and typeset by The Open University, using the Open University TEX System.

Printed and bound in the United Kingdom by The Charlesworth Group, Wakefield.

ISBN 978 0 7492 2870 5

2.1

Contents

Study guide

You have met the idea of proportion in earlier units and Section 1 consolidates and builds on this previous experience. This section introduces different types of proportion; it is important that you understand these so that you can build on them in the rest of the unit.

Section 2 uses the *Calculator Book*; you will encounter another form of regression: power regression. You may want to look back at earlier chapters of the *Calculator Book* concerned with other forms of regression, if you find you have forgotten some of the techniques.

It is better (although not essential) that you study Section 2 before Section 3, which uses a video band to illustrate both the use of the proportionality concepts from Section 1 and the power regression of Section 2 in a practical context.

Section 4 draws together many mathematical ideas from Blocks B and C, summarizing and extending the library of functions. It is a long section, as it consolidates much of your mathematical work from two blocks, so you may prefer to study it over several study sessions. You should try to combine your Handbook entries from the units in both Blocks B and C during your study of this section. The regression facilities on your calculator are also summarized in this section.

Section 5 then uses functions from this library in a number of modelling situations.

There are two television programmes which are relevant to this unit: *Designer Rides* introduces some of the ideas about motion which are relevant in Section 5, and *Deadly Quarrels* uses the concepts of proportionality and rates of change, discussed in this unit, in models of warfare and the arms race.

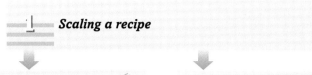 **Scaling a recipe**

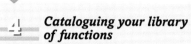 **Finding a power law or proportional relationship from data**   **Experimenting with fruit cake**

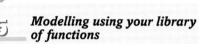 **Cataloguing your library of functions**

Modelling using your library of functions

Introduction

In *Unit 6* on maps, you saw that the measured distance between two points on a map can be scaled up, by multiplying by a suitable scale factor, to give the distance on the ground between the two places represented by the points on a map. You may also remember that to scale up an area you had to multiply by this same scale factor squared.

In *Unit 9*, you saw that the respective frequencies of any two adjacent notes in a Western equal temperament scale were related by the same multiplicative scale factor ($\sqrt[12]{2}$).

In your everyday life, you probably scale things up or down quite often, though you may not be explicitly aware of the different ways in which you do this. This unit uses the context of food, and in particular the scaling of a recipe for making a cake as an example, to help you to appreciate different types of proportionality. However, the mathematics introduced has applications in a very wide variety of contexts, so try to think of other contexts in your own life where similar scaling and ideas of proportionality are appropriate.

Throughout this course, you have been encouraged to be aware of, and consciously monitor your own study and progress as well as integrate your use of the different skills you are developing. In this unit, you are asked to put all your learning experience into practice as you consolidate the mathematics of Blocks B and C. Regularly, try to 'stand back' from what you are doing and analyse both the underlying mathematics and your learning of mathematics. Activity 1 asks you to think how you might go about this.

As usual, there is a Handbook sheet associated with this unit; add your notes to it at appropriate points.

Activity 1 *Monitoring and reviewing learning*

At this stage, you have considerable experience of learning with MU120 and you have been introduced to a variety of skills and ways of working. As you have studied the units in this block, you have been developing a range of skills and techniques that may be useful to you in thinking about how you are doing and learning mathematics.

In working through this unit, try consciously to use some of the learning techniques that you have been introduced to so that you can evaluate their usefulness. The techniques include:

◇   identifying what you can do when you feel stuck;

◇   seeing, saying and recording to help your learning;

◇   using the modelling cycle to identify where you are in a problem.

At particular points, review how you are using the techniques and how you are getting on by asking yourself questions, such as:

◇ What am I doing?

◇ How am I doing?

◇ How well am I doing?

◇ Does this make sense?

◇ At which points might I find it useful to spend time reviewing my work?

1 Scaling a recipe

Aims   This section aims to introduce different types of proportionality, the associated mathematical relationships, and their graphs.   ◇

Recipes for dishes are often given in recipe books for a particular number of servings or for a specific size of tin.

▶   What do people do if they want a different number of portions or they do not have a tin this size?

This section looks at the common problem of scaling a recipe, and highlights a very important mathematical concept—that of proportion—which applies in an enormous number of other contexts. The simplest type of proportional relationship leads to a linear function represented by a straight-line graph. Other types of proportional relationship lead to different functions and their associated graphs, which add to the library of functions that you can use in mathematical modelling.

1.1 Ingredients for different numbers of people

Suppose that you were catering for five people for a meal and you had decided to try to cook them caramel cream for dessert. However, your cookery-book contains only the following recipe, which you would need to adapt.

Caramel cream (serves 3)

90 g brown sugar
45 g castor sugar
$\frac{1}{2}$ litre full cream milk
2 large eggs
1 teaspoon of vanilla essence

Caramelize the brown sugar in a thick saucepan over a low heat until it is dark brown. Pour quickly into an oven-proof dish, spreading the liquid caramel evenly over the dish by tipping it. Heat the milk with the vanilla. Beat the eggs with the castor sugar, then add the warm milk slowly to the eggs and sugar. Whisk the mixture and pour over the caramel. Stand the dish in a baking tin of water and bake at mark 2 (105 °C) for $1\frac{1}{4}$ hours.

▶   How might you adapt the recipe for five people?

You might say that it is easier to cater for six people and give people a slightly bigger helping: simply double the number of servings and double the ingredients. However, you might be more thrifty or anxious to do it more precisely, and decide to bake just five servings, multiplying all the ingredients by the same amount, $\frac{5}{3}$. This number acts as a multiplicative scale factor in the same sense as was used in *Units 6* and *9*.

Suppose that you took this latter course of action.

▶   Which numbers in the recipe would you need to change and how would you change them?

You would need to change the quantities of all the ingredients:

90 g brown sugar would become $\frac{5}{3} \times 90$ g, which is 150 g;

45 g castor sugar would become $\frac{5}{3} \times 45$ g which is 75 g;

the milk would become $\frac{5}{3} \times \frac{1}{2}$ l which is $\frac{5}{6}$ l or nearly 1 litre;

the vanilla essence would become $\frac{5}{3}$ teaspoons or nearly 2 teaspoons;

scaling the eggs by $\frac{5}{3}$ gives $\frac{10}{3}$ or $3\frac{1}{3}$ large eggs, but this is not very practicable for a recipe and so you might decide to use 3 extra large eggs or 4 standard size eggs.

You might also need to cook the caramel cream for a little longer than $1\frac{1}{4}$ hours, but the cooking temperature would be unlikely to need changing.

Apart from the rounding for convenience, you would scale the ingredients by a factor of $\frac{5}{3}$, or equivalently divide by 3 and multiply by 5.

Generalizing this procedure to one which would be applicable for scaling the ingredients in any recipe for a different number of servings gives the following ordered sequence of instructions, called an *algorithm*.

(a) Divide all the ingredients by the number of servings suggested in the recipe to get the amount for one serving.

(b) Multiply the amount for one serving by the number of servings required.

(c) Round the amounts appropriately.

▶   Suppose that you had decided to adapt the recipe for six portions; that is, double the number in the recipe: does the above procedure double all the ingredients too?

Yes. It comes down to dividing by 3 and then multiplying by 6, which is the same as doubling.

The idea of scaling a recipe provides a particular example of the mathematical concept of proportion. The amount of each ingredient required is the amount for one portion (or serving) multiplied by the number of portions. Put another way, the amount required of a particular ingredient a is equal to the number of portions p multiplied by the amount of that ingredient required for one portion k (a constant). So the relationship can be written symbolically as:

$$a = kp \tag{1}$$

The amount a of a particular ingredient is said to be *directly proportional* to the number of portions p. This can also be expressed as 'a and p are directly proportional' or 'there is a direct proportional relationship between a and p'. The word 'directly' is used to distinguish this from some other types of proportional relationship, which you will meet later in this unit.

Mathematically, one variable (a in this case) is said to be directly proportional to another (p in this case) if when one doubles (or trebles, or halves, or is multiplied by $\frac{5}{3}, \ldots$), so does the other. Double p, the number of portions, and the result is to double a, the amount of the ingredient.

You have met this idea before in *Units 2, 6* and *7.*

Another piece of mathematical language describes the constant k in (1) as the *constant of proportionality*, in this case, the amount for one serving. In a direct proportional relationship one variable equals a constant (the constant of proportionality) times the other variable.

There is a mathematical symbol which is shorthand for directly proportional: it is $\propto$, and so instead of the sentence 'the amount of an ingredient is directly proportional to the number of servings', you can write

amount of ingredient $\propto$ number of servings

or shorter still:

$$a \propto p \tag{2}$$

Another way to write (1) is to divide both sides by p, giving:

$$\frac{a}{p} = k \tag{3}$$

The ratio language may help you to connect with ideas from *Unit 9.*

This says that $\frac{a}{p}$ is always a constant or, evidently, that the ratio $a : p$ is constant. Any of (1), (2) or (3) can be used to indicate a direct proportional relationship. They are equivalent representations of the same relationship.

Look at (1) and think about whether you have met this type of relationship before.

▶   What sort of graph would it give?

It is a linear relationship:

$$a = kp$$

This corresponds to $y = mx + c$, with the variables y and x replaced by a and p, the gradient m is k in this case, and the intercept c is zero. So the graph of this relationship (plotting a against p for a given value of k) is a straight line, whose slope (or gradient) is the constant of proportionality k; it has a zero intercept, and so the line goes through the origin: see Figure 1.

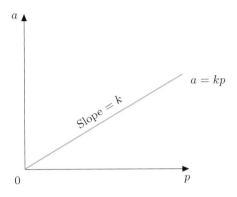

Figure 1

Looking at the context of the ingredients of the caramel cream recipe, the quantity of brown sugar (in grams) is 30 times the number of servings, so the constant of proportionality in this case is 30 (the number of grams for one serving). So:

$a = 30p$

For castor sugar, the constant is 15 (the number of grams for one serving). So the relationship is:

$a = 15p$

The graphs of these relationships are given in Figure 2.

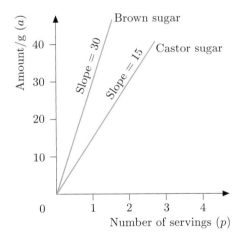

Figure 2

Activity 2   *Scaling the other ingredients*

What are the constants of proportionality for the other ingredients: milk, eggs and vanilla essence? Write down algebraic expressions specifying these proportional relationships and sketch their graphs.

Any direct proportional relationship can be represented by a linear equation like $a = kp$ and by a straight-line (or linear) graph, passing through the origin, as in Figure 1. Sometimes direct proportional relationships are referred to as linear proportional relationships.

A direct proportional relationship can be represented in a number of alternative but equivalent forms. It can be represented by a proportional relationship:

$$y \propto x$$

Or it can be represented by a linear function:

$$y = kx$$

Or it can be represented by a straight-line graph, through the origin as shown in Figure 3.

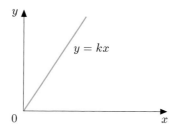

Figure 3

Activity 3   *Cooking a fruit cake*

Suppose you went to a large birthday party and had a piece of a birthday cake you liked so much that you asked for the recipe. The friend who baked it kindly obliged, but the recipe was for a 20 lb cake, and you would like to adapt it to make smaller cakes for special occasions.

'lb' is an abbreviation for the pound, the imperial unit of mass. $1\,\text{lb} \simeq 0.45\,\text{kg}$.

Cake recipe (20 lb cake)

Mixed fruit	7 lbs
Glacé cherries	1 lb
Flaked almonds	1 lb
Mixed peel	1 lb
Plain flour	3 lbs
Mixed spice	5 teaspoons
Butter	$2\frac{1}{2}$ lbs
Brown sugar	$2\frac{1}{2}$ lbs
Eggs (beaten)	18
Brandy	6 tablespoons
Cooking time	10 hours
Cooking temperature	300 °F (150 °C)
Side of square cake tin	25 cm (10 inches)
Height of cake	11 cm (4.5 inches)
Diameter of circular cake tin	30 cm (12 inches)
Height of cake	10 cm (4 inches)

(a) The proportional relationships for each ingredient, which would enable you to scale down the recipe to any given weight of cake, could all be written in the form $a = kp$. Write down the particular value of k for each ingredient. (Assume you are making a 1 lb cake.)

(b) Use your answers to part (a) to scale down the ingredients for this cake recipe, first to a 5 lb one and then to a 7 lb one.

Direct proportional relationships occur frequently as assumptions in mathematical models of everyday situations. Some further examples are given below.

(a) In given traffic conditions, the petrol consumed by a particular car will be approximately directly proportional to the distance driven. The constant of proportionality is the rate of petrol consumption (in litres per mile or per kilometre, say).

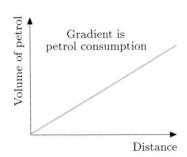

(b) The monthly interest charged on a credit-card account is proportional to the debit balance at the start of the month (assuming no cash advances are made). The constant of proportionality is the monthly rate of interest.

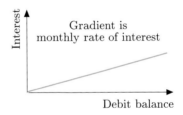

(c) For a wild animal population, which is small in comparison with the numbers which its habitat could sustain, the annual number of births will be approximately proportional to the size of the population. The constant of proportionality is the annual birth-rate per head of the existing population.

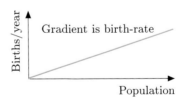

(d) The volume of paint required to cover the walls of a room is proportional to the area of the walls to be painted. The constant of proportionality is the average thickness of the paint.

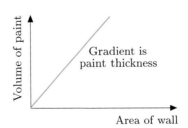

(e) In a lump of matter containing carbon-14, the number of atoms of carbon-14 decaying in a year is proportional to the number of carbon-14 atoms present at the beginning of the year. The constant of proportionality is the radioactive decay constant for carbon-14.

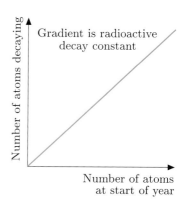

Example 1   *Water consumption*

In a village where new houses are being built quite quickly, the local water company needs to predict the way in which water consumption will increase with increased population. It assumes that water consumption is directly proportional to the population; that the current population is 2000; and the current water consumption is 250 cubic metres per day.

Produce a formula which will predict the water consumption (w cubic metres per day) for a given population (P people). Use it to predict the water consumption after new housing estates are built which are predicted to cause the population to rise first to 3000, then to 4500 and finally to 5500.

Assume that $w \propto P$ or, equivalently, if the water consumption per person is $k\,\text{m}^3$ per day, that:

$$w = kP \tag{4}$$

or equivalently that:

$$\frac{w}{P} = k \tag{5}$$

Substitute the given data into (5) to find k.

When $P = 2000$, $w = 250$ and so:

$$k = 250/2000 = 0.125$$

Substituting this value of k in (4) gives the required formula:

$$w = 0.125P$$

When $P = 3000$, $w = 0.125 \times 3000 = 375\,\text{m}^3$.
When $P = 4500$, $w = 0.125 \times 4500 = 562.5\,\text{m}^3$.
When $P = 5500$, $w = 0.125 \times 5500 = 687.5\,\text{m}^3$.

So the model predicts water consumption of 375, 562.5 and 687.5 m^3 per day for populations of 3000, 4500, and 5500 people respectively.

1.2   Ingredients for a different size of tin

The recipe for the large birthday cake in Activity 3 suggested the sizes of the cake tin to use for the 20 lb cake as either a 10-inch square one with a height of at least $4\frac{1}{2}$ inches or a 12-inch diameter tin with a height of at least 4 inches. Suppose that you wanted to scale down the recipe to fit the sizes of cake tin which you already had: a 5-inch square tin and a 7-inch diameter round tin, both tins being $2\frac{1}{2}$ inches high.

▶   How could you do it?

This relationship is not as simple as direct proportion, because the weight of the cake will be directly proportional to its *volume*, not to the diameter, height or length of the cake tin. You will need to find the relationship between the dimensions of the tin and its volume.

Remember that mass is often referred to colloquially as weight.

The 5-inch square tin

For the 5-inch square tin, you can halve all the dimensions of the 10-inch square cake, and bake a 5-inch square cake of height $2\frac{1}{4}$ inches (just less than the height of the tin!), but you do not halve the amount of the ingredients.

> The volume of the 10-inch square cake would be $10 \times 10 \times 4\frac{1}{2}$ or 450 cubic inches.
>
> The volume of the 5-inch square cake would be $5 \times 5 \times 2\frac{1}{4}$ or 56.25 cubic inches.

The ratio of these volumes is 450/56.25, which is 8. So although the dimensions of the big cake are twice those of the small one, its volume is eight times that of the small one. Since both cakes would be made of the same mixture, the amounts of the ingredients for the big cake would be eight times those of the smaller one.

This may be a rather surprising result, but it is actually true more generally—double the dimensions of any solid and you do not double the volume, but increase it by a factor of 8 (which is 2 cubed). So in order to adapt the recipe for the 5-inch square tin, the ingredients for the 10-inch square cake need to be divided by 8 (which is the same as *multiplying by $\frac{1}{8}$*): halving all the dimensions means you scale the volume by $\frac{1}{8}$, which is $\left(\frac{1}{2}\right)^3$.

This result can be generalized to all cakes in the shape of a cuboid.

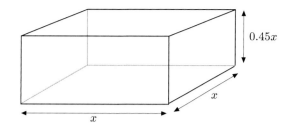

Figure 4   A square-based cake

Look at Figure 4. The cake has dimensions x, x and $0.45x$. So its volume is $0.45x^3$. If the dimensions are doubled, they will become $2x$, $2x$, and $0.9x$, so the volume will be $2x \times 2x \times 0.9x$, which is $3.6x^3$ which equals $8 \times 0.45x^3$ (eight times the volume of the smaller tin). If the dimensions were halved, they would be $\frac{x}{2}$, $\frac{x}{2}$, and $\frac{0.45x}{2}$, and so the volume would be $(\frac{1}{2})^3 0.45x^3$.

So the ingredients of the original recipe would have to be scaled by a factor of $\frac{1}{8}$ for the 5-inch square tin, and so the weight of the cake would also be scaled by $\frac{1}{8}$ resulting in a $\frac{20}{8} = 2.5\,\text{lb}$ cake.

The 7-inch diameter tin

The same recipe was used for a 12-inch diameter tin and so, for the 7-inch tin, you would scale all the dimensions of the big circular cake by a factor of $\frac{7}{12}$. This gives a height of $4 \times \frac{7}{12} = 2\frac{1}{3}$, which should fit the $2\frac{1}{2}$-inch high cake tin.

▶   What would this do to the volume of the cake?

The cake recipe gives the dimensions of a circular cake showing that the height of this particular cake is one-third of its diameter, or two-thirds of its radius, as shown in Figure 5.

Mathematically, this shape is called a *cylinder* and the volume of any cylinder is $\pi \times (\text{radius})^2 \times \text{height}$.

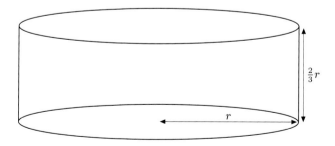

Figure 5   A circular-based cake

So for this particular cake of radius r and height $\frac{2}{3}r$, the volume is $\pi r^2 \times \frac{2}{3}r = \frac{2}{3}\pi r^3$.

Activity 4 *Scale volume*

If all the dimensions of a cake are scaled by a factor of $\frac{7}{12}$, by what factor is the volume scaled?

The volume is scaled by a factor of about 0.2 (or 1/5), and so the amount of each ingredient will need to be scaled by this factor for the 7-inch diameter cake. That is, the amount of each ingredient will need to be multiplied by 1/5, which is the same as being divided by 5.

Activity 5 *The x cake*

Suppose you had the recipe for a cake in the shape of a cube, with sides of length x. Find the volume of the cake that you would make if you were to:

(a) treble all the dimensions;

(b) divide all the dimensions by three.

For any shape of cake, if all the dimensions are scaled by the same factor, then the volume is scaled by the cube of this factor. This is called a *cubic proportional* relationship.

There is another way of expressing this type of relationship. Look at the following table.

Table 1 Cake shape, dimension and volume

Cake shape	Cake dimension	Cake volume
Cube	side x	$V = x^3$
Square-based cake	square side x height $0.45x$	$V = 0.45x^3$
Circular-based cake	radius r height $\frac{2}{3}r$	$V = \frac{2}{3}\pi r^3$

In each case, the volume is the cube of one dimension—(dimension)^3—multiplied by a constant. In the case of the cube, the constant is 1; for the square-based cake, the constant is 0.45; in the case of the circular-based cake, it is $\frac{2}{3}\pi$.

This means that the volume is directly proportional to the *cube* of one of its dimensions, x say. So:

$$V \propto x^3$$

The constant of proportionality depends on the shape and the dimension chosen. For example, if the diameter of the circular-based cake had been given instead of the radius, then the particular constant of proportionality would be different. The volume would be $\frac{2}{3}\pi(d/2)^3 = \frac{2}{3}\pi d^3/8 = \frac{1}{12}\pi d^3$. So the constant of proportionality would be $\frac{1}{12}\pi$.

If you scale up all the dimensions of any three-dimensional solid by the same factor, you scale the volume by the cube of that factor.

In general, if you scale all the dimensions of a container, like a cake tin, by a particular scale factor, you scale the volume of its contents by the cube of this scale factor. This has many other applications in everyday life. For instance, when shopping, and comparing the prices of different-sized containers of the same product, it is important to compare the volumes or weights not the heights or other dimension of the containers.

Example 2   Egg sizes

Many recipes include the use of eggs, and earlier you saw that in order to avoid 'fractional' eggs after scaling a recipe, you might use different-sized eggs: for example, instead of $3\frac{1}{2}$ eggs, use 4 smaller eggs. You could be more precise about how much smaller the eggs should be in order to get the same amount of egg. To do this, you need to know how the amount of egg varies with UK egg size.

Imagine that you wanted to investigate this question for yourself before buying the eggs; so you took a ruler into a shop and measured the length of eggs of different sizes of eggs and obtained the results in the table below.

UK egg size	1	2	3	4
Length/mm	67	65	62	59

(a) How is the volume of each size of egg related to the volume of a size 1 egg, assuming that all the eggs are precisely similar in shape; that is, scaled versions of each other?

The idea of similarity was mentioned in *Unit 2*. It also forms a major focus of the next unit.

(b) How many whole eggs of different sizes would be equivalent to $3\frac{1}{2}$ size 3 eggs?

To answer part (a), assume that all the dimensions of eggs of different sizes are scaled up by the same factor. This leads to assuming that the volume of an egg is proportional to the cube of its length. If this is the case, then the relationship is given by:

$$V \propto l^3$$

So the volumes of the different grades of egg compared with a size 1 egg will be the ratio of the cubes of their lengths compared with the cube of the length of a size 1 egg. Hence you need to work out these cubed values. Notice that this gets round the complex problem of finding the actual volume of an egg. The values are given in the table below.

UK egg size	1	2	3	4
Length/mm	67	65	62	59
(Length/mm)^3	300763	274625	238328	205379
Volume in terms of one size 1 egg	1	$\dfrac{274625}{300763}$ $\simeq 0.91$	$\dfrac{238328}{300763}$ $\simeq 0.79$	$\dfrac{205379}{300763}$ $\simeq 0.68$

To answer part (b), you need to find out what is equivalent to $3\frac{1}{2}$ size 3 eggs. So you need to find its equivalent in size 1 eggs.

$$3\tfrac{1}{2} \times 0.79 = 2.8 \text{ size 1 eggs}$$

Three size 2 eggs (equivalent to 3×0.91 or 2.7 size 1 eggs) or four size 4 eggs (equivalent to $4 \times 0.68 = 2.7$ size 1 eggs) would be reasonably close.

Cubic proportional relationships can be symbolized by, for example, $V \propto r^3$ or specified by a function like $V = 0.45r^3$. They can also be represented graphically.

The graphs of cubic proportional relationships are not straight lines. Use your calculator to plot (for positive x) the ones you have met so far, namely:

$$y = x^3 \qquad y = 0.45x^3 \qquad y = \tfrac{2}{3}\pi x^3$$

Check your graphs against Figure 6.

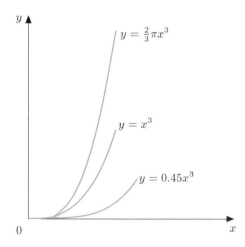

Figure 6   Cubic graphs

The graphs all share some common characteristics: y increases as x increases and the larger the value of x, the steeper the slope (the larger the numerical value of the gradient). This is what is called a *concave* curve. If you have not met the term 'concave' before, then you may find it helpful to think of it as part of a smile as in Figure 7(a), although the dictionary definition is 'curving inwards'. *Convex* curves are like part of a 'vexed' scowl, as in Figure 7(b), or according to the dictionary they 'curve outwards'.

Figure 7   (a) concave curves (smiling)   (b) convex curves (scowling)

▶   Think about what the words 'inwards' and 'outwards' from the dictionary definition mean mathematically.

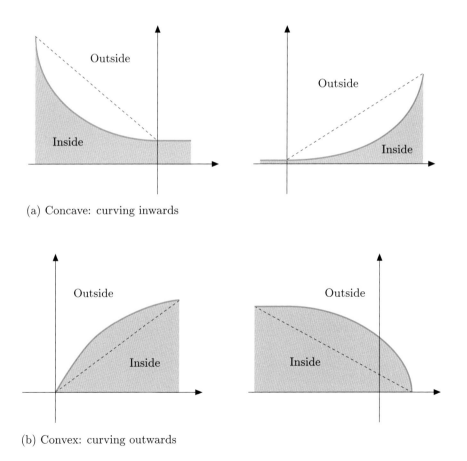

(a) Concave: curving inwards

(b) Convex: curving outwards

Figure 8   Inside and outside a graph

In the context of graphs, it would seem that you could regard the area under the graph as 'inside' and the area above the graph as 'outside' the curve; if so, then the dictionary definition fits very well.

Notice that all the graphs of cubic proportional relationships (see Figure 6) pass through the origin, as did the graphs of the direct (linear) proportional relationships.

So cubic proportional relationships can be specified by:

$$y = kx^3 \qquad \text{(where } k \text{ is the constant of proportionality)}$$

or symbolized by:

$$y \propto x^3$$

or shown by a cubic (*concave*) graph through the origin.

1.3 Icing for different-sized cakes

Once you have made a cake, you might like to ice it. If you had scaled a recipe for the cake, you would also need to scale the quantity of icing for the top surface.

▶   How would the area of the top surface of the cake be scaled?

First, consider the square-based cake. The area A of the top of a square-based cake, as in Figure 4, of side x is just x^2.

So $A = x^2$.

However, if you double the side of the square, to $2x$, the surface area of the top is:

$$A = (2x)^2 = 4x^2$$

So doubling the dimensions of the cake has scaled the area to be iced by 2^2 or 4.

The area of the top of the cake is proportional to the square of the side.

$$\text{Area} \propto (\text{side})^2$$

The top of a circular-based cake of radius r, as in Figure 5, will have a surface area A of πr^2:

$$A = \pi r^2$$

So if you doubled the radius, the surface area of the top of the cake would be scaled by 2^2 or 4. The area of the top is proportional to the square of the radius of the cake.

$$\text{Area} \propto (\text{radius})^2$$

So the area of both shapes of cake is proportional to the square of one of its dimensions. This type of relationship is called *square proportional*.

$$\text{Area} \propto (\text{dimension})^2$$

More generally, if all the dimensions of a given shape are scaled by the same factor, then any surface area of the shape is proportional to the square of one dimension of the shape.

Example 3

Suppose that you have scaled a cake recipe from a 12-inch diameter cake tin to a 7-inch diameter cake tin, by scaling the ingredients by $(7/12)^3$ or about one-fifth. You want to follow the recipe and put almond paste and icing on both the top and the sides of the cake. Find the factor by which you should scale these ingredients.

You will have scaled all the dimensions of the cake by 7/12, so the area of the top will be scaled by $(7/12)^2$.

Similarly, the area of the side of the cake would be scaled by $(7/12)^2$. So the almond paste and icing ingredients should be scaled by a factor of $(7/12)^2 \simeq 0.3402782$, which is about one-third.

Square proportional relationships can be represented by graphs. The functions describing square proportional relationships for the area of the top of a cake are:

$$y = x^2 \quad \text{(for a square-based cake)}$$
$$y = \pi x^2 \quad \text{(for a circular-based cake)}$$

The graphs of these functions both give parabolas, passing through the origin, as shown in Figure 9.

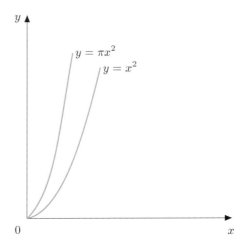

Figure 9   Graphs of square proportional relationships

So *square proportional* relationships can be represented by:

$$y = kx^2$$

where k is the constant of proportionality;

or symbolized as:

$$y \propto x^2$$

or shown by the graph of a parabola, passing through the origin.

1.4 Many hands make light work

So far you have been considering the effects of changing the number of portions for a recipe or the size of tin on the quantity of ingredients used; these lead to various sorts of proportional relationships, where if the number of portions or the sizes of the tins go up so do the quantity of each ingredient. However, sometimes the opposite type of relationship holds. For instance, if you have more people to help to do something, it will usually take less time, assuming of course that you subscribe to the philosophy of 'many hands make light work' rather than 'too many cooks spoil the broth'.

Suppose you are catering for a party for a large number of people and there are many jobs to be done: food to prepare and tables to be laid. In general, if there are two people working to get all the jobs done, it should take about half the time. If there are three, it should take about one-third the time; if four, one-quarter the time; and so on. This type of relationship is called an *inverse proportional relationship*.

Suppose that you estimate that all the jobs would take one person 12 hours, then two people should take 6 hours, three people should take 4 hours, and four people 3 hours. The time taken is said to be inversely proportional to the number of people, (up to a certain point when too many additional people could get in each other's way). You could write this relationship as:

(number of people) $\times$ (time taken) $= 12$ person-hours

or

$$\text{time taken} = \frac{12 \text{ person-hours}}{\text{number of people}}$$

If N is the number of people, and T the time taken in hours, then these relationships become:

$$NT = 12$$

or

$$T = \frac{12}{N}$$

In this case, you can say that T is *inversely proportional* to N, the constant of proportionality being 12, and you could write:

$$T \propto \frac{1}{N}$$

The latter is a general statement and can be read as 'T is directly proportional to the reciprocal of N', or 'T is inversely proportional to N'.

This type of relationship would hold in other workload situations where the time for one person to do all the jobs differed from 12 hours. However, in every such situation, there would be a range for the number of people for which the 'many hands make light work' model would be applicable. You cannot subdivide a list of jobs indefinitely and some may depend on others being completed—at some point you would need a different model: perhaps 'too many cooks spoil the broth'.

$T \propto \dfrac{1}{N}$ is an example of an inverse proportional relationship, which can also be represented by:

$$T = \frac{k}{N}, \text{ where } k \text{ is the constant of proportionality,}$$

or by a graph.

▶ The graphs of the inverse proportional relationships: $y = 1/x$; $y = 2/x$; and $y = 3/x$ are shown below.

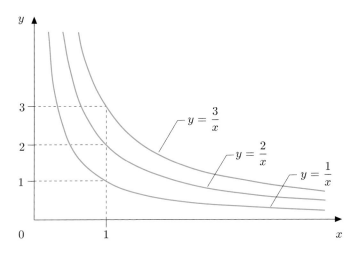

Figure 10 Graphs of inverse proportional relationships

Once again, the particular units of measurement have been ignored, and the mathematical relationships between the variables stressed.

Inverse proportional relationships occur in modelling many contexts. For instance, when you are travelling a fixed distance, d say, the faster your average speed s, then the shorter the time t of your journey. In fact:

$$s = \frac{d}{t}, \quad \text{for a given distance } d$$

This is an inverse proportional relationship, one between the average speed s and the journey time t. The relationship can be rearranged to give:

$$t = \frac{d}{s}, \quad \text{where } d \text{ is a constant}$$

So another way of expressing the relationship is to say that the journey time t is inversely proportional to the average speed s with the same constant of proportionality being the distance, d. So $t \propto \dfrac{1}{s}$ for a constant distance.

Just as direct proportional relationships can be square proportional or cubic proportional, so can inverse proportional relationships.

Example 4

Suppose you had all the ingredients mixed for a cake recipe, when you noticed that the recipe was for a 5-inch square cake tin, but you only had a 4-inch or a 7-inch square tin. So you wondered what would happen to the height of the cake if you used one of these tins, with the amount of ingredients from the original recipe.

The volume V of the cake would be constant, but its dimensions could vary. Consider a square-based cake with the sides of the base x and height h. Then:

$$x^2 \times h = V$$

Divide both sides of this equation by x^2 to get h as the subject of the formula, because it is the height you are interested in. This gives:

$$h = \frac{V}{x^2}$$

So in this case the height is inversely proportional to the square of the side of the base.

If the original height using a 5-inch tin is H, and the new height with a different square-based tin of side x is h, then:

$$H = \frac{V}{5^2} \tag{6}$$

$$h = \frac{V}{x^2} \tag{7}$$

In order to get H in terms of h, you need to eliminate V from equation (7). This can be done by manipulating equation (6) to find V in terms of H, and then substituting this expression for V in equation (7):

$$V = 5^2 H$$

Now substitute for V in (7):

$$h = \frac{V}{x^2} \text{ becomes:}$$

$$h = \frac{5^2 H}{x^2}$$

This can be rewritten as:

$$h = \left(\frac{5}{x}\right)^2 H$$

You have a 4-inch and a 7-inch square tin, so $x = 4$ or $x = 7$.

If $x = 4$, then:

$$h = \left(\frac{5}{4}\right)^2 H \simeq 1.5625 H$$

So the height would be increased by over half as much again. You would need to consider whether the tin was deep enough.

If $x = 7$, then:

$$h = \left(\frac{5}{7}\right)^2 H \simeq 0.5102H$$

So the height would be almost halved. You might think twice about using this tin, without adding some extra ingredients.

A small change in the size of the side of the tin produces a more significant change in the height, because the scale factor is squared. The scaling factor for the sides is $7/5$, but the scaling factor for the (height) is:

$$\frac{1}{(\text{scaling factor})^2}$$

if the volume is to be unchanged.

You have now met a couple of different inverse proportional relationships. Now think about the shape of the graphs of such relationships. As one variable increases the other decreases; for instance, double one, halve the other. So the graphs are very different from those for direct proportional relationships.

Use your calculator to plot:

$$y = \frac{1}{x}, \quad y = \frac{1}{x^2} \quad \text{and} \quad y = \frac{1}{\sqrt{x}}$$

Check your graphs with those in Figure 11.

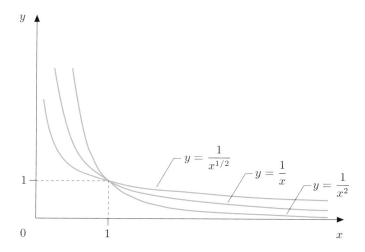

Figure 11   Graphs of different inverse proportional relationships

1.5 Equivalent proportional relationships

Proportional relationships can be turned around and expressed in equivalent forms for different purposes. This subsection looks at how you can do this, as well as at the general form of all proportional relationships. This form is particularly useful when trying to decide which, if any, proportional relationship is appropriate for experimental data.

The weight of a cake W (made of a specified mixture) will be proportional to its volume V, but you could equally well say that the volume V was proportional to the weight W of the cake. Both ways of expressing the relationship may be useful. Symbolically, this can be written as:

if $W \propto V$, then $V \propto W$; and

if $V \propto W$, then $W \propto V$

Another example of turning a proportional relationship round involves travelling at constant speed. If you travel at a constant speed, then the distance d that you travel will be proportional to the travel time t. This can be written as:

$d \propto t$ or $d = st$ (where s is the constant speed)

So the constant of proportionality is the constant speed s. However, this relationship can be turned round, because the travel time will also be proportional to the distance travelled:

$t \propto d$

This can be seen by manipulating the equation:

$d = st$

to get t as the subject, namely:

$t = \dfrac{1}{s}d$

So t is proportional to d, the constant of proportionality being the reciprocal of the constant speed.

Other proportional relationships can be expressed in different forms, by manipulating the associated equation.

Example 5

The volume V of a square-based cake, of side X, is proportional to the cube of its side X. In symbolic form, this is $V \propto X^3$, or in the form of an equation:

$$V = kX^3 \text{ (where } k \text{ is some constant)} \tag{8}$$

Express this the other way round to find X in terms of V.

This proportional relationship can be expressed the other way round by manipulating the equation to get X as the subject. First divide through by k to get:

$$X^3 = \frac{1}{k}V$$

Then take the cube root to get:

$$X = \left(\frac{1}{k}\right)^{1/3} V^{1/3}$$

Since $\left(\dfrac{1}{k}\right)^{1/3}$ is a constant, this means that X is proportional to the cube root of V.

$$X \propto V^{1/3}$$

This is equivalent to the cubic proportional relationship:

$$V \propto X^3$$

Activity 6   Round and round

The area of the top of a circular-based cake is proportional to the square of the radius. Express this proportional relationship the other way round.

Proportional relationships can also be combined. If the weight W of a round-based cake is proportional to its volume V; that is, $W \propto V$, and the volume V is proportional to the cube of the radius r of the cake; that is, $V \propto r^3$, then it follows that the weight is proportional to the cube of the radius. Symbolically, this is written:

$$W \propto V \text{ and } V \propto r^3 \text{ implies } W \propto r^3$$

Sometimes it is necessary to express a proportional relationship in an alternative equivalent form before combining them.

Example 6   How big, how heavy?

For a circular-based cake of radius r, weight W and top surface area A, the following two proportional relationships hold:

$$W \propto r^3 \tag{9}$$
$$A \propto r^2 \tag{10}$$

How does the area of the top of the cake depend on its weight?

You need to eliminate r from these two proportional relations and get a relationship which gives the area A as proportional to something which involves W but not r.

The strategy is to get a proportional relationship for r in terms of W from (9) and then substitute for r into (10).

Rearranging $W \propto r^3$ gives:

$$r^3 \propto W$$

and taking the cube root gives:

$$r \propto W^{1/3}$$

So $A \propto r^2$ and $r \propto W^{1/3}$.

Hence $A \propto (W^{1/3})^2$ or:

$$A \propto W^{2/3}$$

So if you double the weight of the cake, you scale the area of the top of the cake (that needs to be covered by icing) by $2^{2/3}$ which is $1.587\ldots$ or about one and half times.

Another way to look at the relationship is to take the square root of both sides of the proportionality. So

$$A \propto r^2$$

implies that

$$A^{1/2} \propto r$$

or

$$r \propto A^{1/2}$$

showing that the radius is proportional to the square root of the area of the top of the cake.

Activity 7   *Double helpings*

Suppose that you have a recipe for a 120 mm square-based dish, which you wish to double to provide twice as many servings. You have a 150 mm square-based dish which seems suitable.

(a) Check that the dish is suitable by working out the scale factor.

(b) How would the surface area of the top of the dish change?

(c) If the recipe suggested 100 g of cheese for the top layer, would you need to double it for the doubled version of the dish?

1.6 *Proportion and power laws*

The proportional relationships which you have met are summarized in Table 2, overleaf. They are all of the form:

one variable $\propto$ another variable raised to a power

If the variables are labelled y and x, and the power is n, then it is useful to summarize proportional relationships by either:

$$y \propto x^n$$

or $y = kx^n$ (k is a constant)

The latter is sometimes called a power law or a power relationship. Notice that n could be a positive or negative whole number, or a positive or negative fraction. Some examples are shown in Table 2. Look at it now and think about the value n takes in each case.

Table 2   Proportional relationships, power laws and their graphs

Name of relationship	Proportional relationship	Power law k is constant of proportionality	Graph
Direct or linear proportion	$y \propto x$	$y = kx$	
Square proportion	$y \propto x^2$	$y = kx^2$	
Cubic proportion	$y \propto x^3$	$y = kx^3$	
Square root proportion	$y \propto x^{1/2}$	$y = kx^{1/2}$	
Cube root proportion	$y \propto x^{1/3}$	$y = kx^{1/3}$	

Name of relationship	Proportional relationship	Power law k is constant of proportionality	Graph
Inverse proportion	$y \propto 1/x$	$\begin{aligned} y &= k/x \\ &= kx^{-1} \end{aligned}$	$y = \dfrac{k}{x}$
Inverse square proportion	$y \propto 1/x^2$	$\begin{aligned} y &= k/x^2 \\ &= kx^{-2} \end{aligned}$	$y = \dfrac{k}{x^2}$
Inverse square root proportion	$y \propto 1/x^{1/2}$	$\begin{aligned} y &= k/x^{1/2} \\ &= kx^{-1/2} \end{aligned}$	$y = \dfrac{k}{x^{1/2}}$

When $n = 1$, the relationship is one of direct proportion and the graph is a straight line through the origin.

When $n = 2$ or 3, the relationship is one of square or cubic proportion, and the graph is a concave curve (parabola or cubic) through the origin.

When $n = 4$ or 5, the graphs are similarly concave.

Plot $y = x^n$ for these two values of n and $x \geq 0$, using your calculator.

All these relationships lead to a curve which gets steeper for increasing x. They are all concave like that in Figure 14(a), overleaf.

How about fractional powers of n: that is when n is between 0 and 1? Plot $y = x^n$ for each of $n = \frac{1}{4}, \frac{1}{5}$ using your calculator.

These relationships lead to curves which are convex. They get less steep for increasing x, like the one shown in Figure 14(c).

In general, for n larger than 1 the curve will resemble that in Figure 14(a); for n less than 1 but greater than 0, the curve will resemble that in Figure 14(c); while $n = 1$ gives a straight line, Figure 14(b).

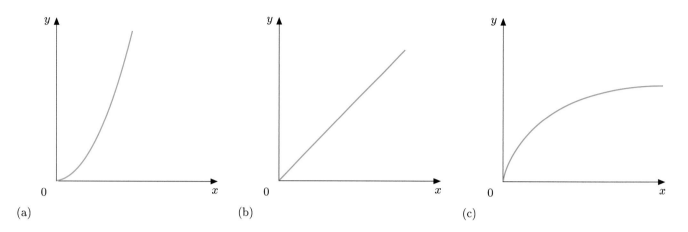

(a)   (b)   (c)

Figure 14

▶   Which function will correspond to $n = 0$?

A constant function, as x^0 is 1. So $y = kx^0$ is $y = k$ and is **not** a proportional relationship, as y does not vary when x does.

When n is less than 0 (that is, negative), $y = x^n$ gives the inverse proportional relationships, which you plotted earlier. Look at the graphs in Figure 15.

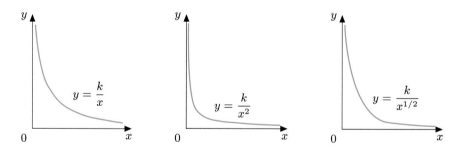

Figure 15

They are all concave graphs, which decrease as they get closer to the x-axis. When a curve approaches a line like this, but never actually touches it, the line is called an *asymptote*. For all these graphs, the y-axis, as well as the x-axis, is an asymptote as the curve approaches it and gets closer and closer to it, but never touches it.

Now you have a number of new functions for your library of mathematical functions to use in modelling. The graphs of these power functions are of several types as shown in Table 3.

Table 3   Graphs of power functions $y = kx^n (x > 0)$

$n > 1$ Increasing concave curve through the origin

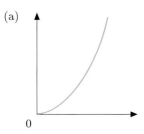

$n = 1$ Straight line through the origin

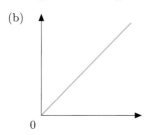

$0 < n < 1$ Increasing convex curve through the origin

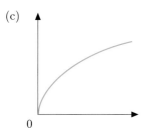

$n < 0$ Decreasing concave curve

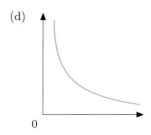

Activity 8 *Power law for circular-based cakes*

The following proportional relationships were all used for the cake. Write down each relationship in the form of a power law and sketch the graphs.

(a) $W \propto V$ (b) $V \propto r^3$ (c) $A \propto r^2$ (d) $A \propto W^{2/3}$

(e) $W \propto r^3$ (f) $r \propto W^{1/3}$ (g) $r \propto A^{1/2}$ (h) $W \propto A^{3/2}$

In this section, you have added a number of additional mathematical functions to your library: proportional or power-law functions, arising from direct or linear proportion, square and cubic proportion, square root proportion, inverse proportion, inverse square proportion. The graphs of these functions are summarized in Tables 2 and 3.

Activity 9 *Add the new functions to your Handbook*

Look back at Tables 2 and 3 and the most important features of proportionality and power-law relationships. Make notes to add to your Handbook sheet; include sketches indicating asymptotes where appropriate.

Scaling
Lengths are scaled by the scale factor.
Areas are scaled by (the scale factor)^2.
Volumes are scaled by (the scale factor)^3.

Outcomes

After studying this section, you should be able to:

◇ use proportional relationships in relevant contexts, for example, to scale recipes for different numbers of servings or different sized containers (Activities 2 to 5, 7);

◇ explain the meaning of, and use correctly, words describing proportional relationships: 'direct or linear proportion', 'square or cubic proportion', 'inverse proportion', 'inverse square proportion' and 'asymptote' (Activity 9);

◇ write down proportional relationships symbolically both in the form of a power law equation and an expression involving the proportion symbol $\propto$ (Activities 6 and 8);

◇ sketch the graph of a given proportional relationship (Activity 8).

2 Finding a power law or proportional relationship from data

Aims   This section explores using the power law regression facility on your calculator to fit a proportional relationship to given data.   ◇

When you have a set of data for which you think a proportional relationship applies, it is often not obvious which particular relationship would be most appropriate from plotting the data. The graphs of square or cubic proportional relationships look very similar, as do inverse proportional relationships to one another. Your calculator has the facility to fit a power law to data and give you the best fit power law for a given set of data, in the same way as it performs linear, quadratic and exponential regression.

2.1 Data to power law

Now study Section 13.1 of Chapter 13 of the Calculator Book.

Interpreting results

The calculator will give the best fit power law for a given set of data. However, data are rarely exact, and so calculator results should be presented and interpreted in this light. There is no point in giving the parameters in the power law to greater accuracy than that provided by the data. The accuracy should also be appropriate to the purpose for the context, and the final result probably expressed only to the accuracy of the least accurate measurement.

Example 7   *Best buy in eggs (1995)*

Table 4

Size	1	2	3	4
Mean mass x grams	75	70	65	60
Price for 6 eggs y pence	87	78	69	65

Table 4 gives data on the prices and masses of different sizes of eggs bought in 1995 in a UK supermarket. What does this mean in terms of which size is the best buy in terms of mass for your money?

Power regression on the calculator gives the best fit relationship between mass and price as a power law of the form:

$$y = ax^b$$

with $a = 0.2700860602$ and $b = 1.334573765$.

From the above relationship, the best fit suggests that $y \propto x^{1.33}$, the graph of which is concave curving upwards. If all sizes of egg were equally good value for money, then a linear relationship would be expected: $y \propto x$. These relationships are both illustrated in Figure 16.

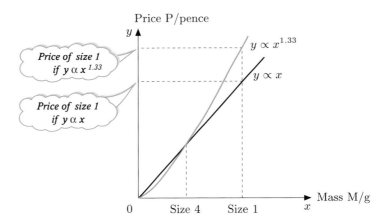

Figure 16 $y \propto x$ and $y \propto x^{1.33}$

The heavier eggs are more expensive and so the smaller eggs appear to be rather better value for money.

▶ But how much would you save by buying six smaller eggs?

Grade 1 eggs are 87 pence for six, each weighing 75 grams, or 0.193 pence per gram (3 d.p.).
Grade 4 eggs are 65 pence for six, each weighing 60 grams, or 0.181 pence per gram (3 d.p.).

The difference is about 7%. This would be quite significant if you were buying eggs in bulk; for instance, if you were in the catering or baking business.

▶ But before you rushed out to buy size 4 eggs in bulk, what other factors might you take into consideration?

This is price per gram of the whole egg and you are probably only interested in the inside of the egg, not its shell. Smaller eggs might have a larger proportion of shell than bigger ones. It is the price per gram of unshelled egg which determines the best buy. Shelling more of the smaller eggs takes more time and if you were paying somebody to do it, the smaller eggs might not be quite such good value.

Testing proportional relationships

Earlier in this unit, when you were considering scaling up a recipe, you considered the number of whole eggs equivalent to $3\frac{1}{2}$ size 3 eggs, by assuming that the volume of an egg was proportional to the cube of its length. This was a reasonable assumption in the absence of much data (the amount you might be able to collect quickly in a shop, before deciding which size eggs to buy). In the case of scaling up a recipe, you would probably not want to invest much time in collecting data on the dimensions of eggs, to test out your assumption. Since the ingredients do not need to be very exact for most recipes, it would be unlikely to affect your culinary product adversely if there were slightly too much or too little egg. However, in other modelling situations, it might be more important to test assumptions. It might, for instance, involve a large cost to a commercial enterprise, or affect the safety of the design of a vehicle or construction. So the following activity is designed to illustrate an important technique rather than being an important modelling exercise in its own right. It is concerned with testing the assumption that egg mass is proportional to the cube of the length of the egg.

Activity 10   *Testing the proportional assumption in Example 2*

The data in Table 5 were collected by a member of the MU120 course team from boxes of six eggs bought in a small UK shop in 1995.

Table 5

UK egg size	1	2	3	4
Mean length/mm	67	65	62	59
Mean diameter/mm	47	45	44	43
Mean mass/g	75	70	65	60

The assumption made earlier in this unit (in Example 2) was that all eggs are of a similar shape, and that the sizes are scaled versions of each other, and so the mass or volume of egg is proportional to its length cubed. Investigate this assumption by looking at the best fit power relationships.

Input the data lists for length, diameter, and mass into your calculator and find the best fit power regression parameters for relationships between the variables: length and mass; diameter and mass; length and diameter. Do these relationships support the assumption that mass is directly proportional to the cube of the length and/or the assumption that eggs of different sizes are scaled versions of each other?

Might this additional data affect the answer to Example 2 on how many whole eggs were equivalent to $3\frac{1}{2}$ size 3 eggs?

Outcomes

After studying this section, you should be able to:

◇   use your calculator to obtain the best fit power law to a given set of data (Activity 10);

◇   use power regression to test proportional relationships (Activity 10);

◇   interpret the results of such a power law fit (Activity 10).

3 Experimenting with fruit cakes

Aims   This section aims to consolidate the skills introduced in Sections 1 and 2 in a practical context.   ◇

This section is based on a video band which features the scaling down of a cake recipe, including the baking time. The main character is Sarah who has been given a recipe by a friend for a fruit cake. Unfortunately, the recipe was for a very large 20 lb cake cooked in a 12-inch diameter cake tin. However, the only suitable cake tin which Sarah has is a 7-inch diameter cake tin. As you watch the video, consider how best to scale the recipe. Before the video begins, Sarah has worked out that scaling all the dimensions of the cake by a factor of 7/12 leads to a scale factor of about one-fifth for the weight of the cake (hence a 4 lb cake). You worked this out in Activity 4. And, in Example 3, you saw that any icing ingredients would need to be scaled by a factor of about 1/3.

3.1 Cooking time

In the end, Sarah decided to experiment with the scaled recipe and did not invest in almond paste and icing for the experiment. However, she did have to scale the baking time.

▶   How would you scale the baking time for a cake or other dish?

Watch the first sequence of the video and note how she did this and what alternatives her guests Judy and Allan suggested.

Now watch the first part of video band 9a of DVD00107 called 'A piece of cake'.

Activity 11   *Cakes again*

What proportional relationship did Sarah originally use for the baking time?

What relationship did Judy suggest from a theoretical analysis of the problem?

What relationship did Allan suggest from his experience?

Sketch the graphs of these three relationships on the same graph.

Now watch the rest of video band 9b of DVD00107.

The experimental results which Allan obtained are given in Table 6 below.

Table 6   Cooking time of cakes of the same shape but different weights

Weight (lb)	1	1.5	2	3	4	6	10	12	16	20
Time (hours)	1.15	1.55	1.80	2.25	2.5	2.8	3.1	3.3	3.6	4

Activity 12   *Best fit*

(a) Enter Allan's experimental data into your calculator and set up a *scatterplot* of cooking time against weight.

(b) Use power regression to find the best fit of the form $y = ax^b$.

(c) Plot a graph of this function on the same screen as the *scatterplot*.

Activity 13   *Empirical and theoretical*

The theoretical model was of the form $y = ax^{1/3}$. Compare this with the best fit regression model by plotting the following functions on the same screen as those for Activity 12.

Use the same value for a as you found in Activity 12 and plot this theoretical model.

Repeat with $a = 1.5$ and compare how well the three functions fit the experimental data.

Activity 14   *A generalized cake tin*

Consider a cake tin of radius r. Use the theoretical model from Activity 13 to write down the proportional relationships for weight, radius and cooking time of similarly shaped cakes. Combine these to give a proportional relationship between cooking time T and radius of cake r.

If you had used the regression model given by your calculator in Activity 12, would this relationship have been simpler or more complicated? So which model is preferable in this case?

Activity 15   *Writing recommendations*

The previous activity considered only the radius of a cake tin. Often, however, diameters are measured. What if you want to scale up the recipe for square-based tins?

(a) Consider a circular-based tin of diameter d. What would the proportional relationship be between the cooking time and the diameter?

(b) Consider a square-based tin of side x. What would the proportional relationship be between the cooking time and the length of the side of the tin?

(c) Write a short paragraph for a recipe book telling people how to scale a recipe by size of tin for both ingredients and cooking time, assuming that all the dimensions of the tin are scaled by the same factor.

(d) Pause to review your learning, and to think about how you have used the calculator and the video sequence. How well do you feel you are working with the idea of proportion and the learning tools mentioned in Activity 1?

In this section, you have seen an example of using proportional relationships to obtain a theoretical model for a practical problem. This model was tested experimentally, the data entered into a calculator, and the best fit power law obtained for comparison with the theoretical model.

The results of this were then used to devise a model for how cooking time varies with size of tin, as well as weight, in such a way that it could be used as a description in a recipe book.

Warning

This proportional relationship between cooking time and radius fits the data well for the range of fruit cakes between 1 and 20 lb of the shape shown in the video. However, extrapolating much beyond these values is inadvisable.

Fruit cakes are cooked slowly in a warm oven. The model might not be applicable to different dishes which are cooked quickly in a hot oven. In such cases, the time for the heat to be absorbed by the surface and to penetrate into the centre of the dish might also need to be taken into account.

Outcomes

After studying this section, you should be able to:

⬦ transfer the skills and techniques of Sections 1 and 2, regarding proportional relationships and power law regression to practical contexts (Activities 11 to 15).

4 Cataloguing your library of functions

Aims   This section aims to summarise and extend your knowledge of the mathematical functions which you have met in the last two blocks, and which may be of use to you in modelling. As such, you should aim to consolidate and add to the notes in your handbook on the relevant mathematical functions while studying this section.   ◇

This section looks at a number of types of mathematical functions and techniques for your library of functions. You have met most of the functions discussed before. Each subsection deals with a different type of function and includes a handbook activity. It will also provide you with an opportunity to monitor and review your learning about the various functions and aspects of mathematical modelling. This section also will serve as a block summary—to this end there are a number of consolidation Learning File activities. The section is quite long and so you may like to split it over more than one study session.

4.1 Linear functions

These functions generate straight-line graphs. They are often used in modelling over short ranges where a linear approximation is reasonable. Because they are simpler mathematically than most other functions, they may often be used as a first model which might be refined later.

Remember that the equation specifying a linear function is of the form:

$$y = mx + c$$

The number m is the gradient or slope of the line, which is also the rate of change of y as x changes. The number c is the y-intercept or the 'starting value'; that is, the value of y when x is zero, where the line crosses the y-axis.

There are two special cases of linear functions.

The first, shown in Figure 17, is when the gradient m is zero, so y is a constant function and the graph is a straight line parallel to the x-axis.

$$y = c$$

The second special case shown in Figure 18 is when the starting value c is zero, which leads to a straight-line graph passing through the origin.

$$y = mx$$

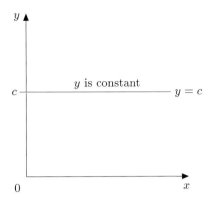

Figure 17

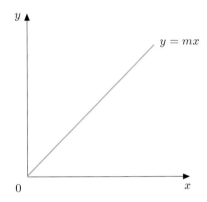

Figure 18

This latter type of function also produces a direct proportional relationship. So in this case y is proportional to x, the constant of proportionality being the value m. This can be written as:

$$y \propto x$$

Often direct proportional relationships are only valid for positive m and when both variables are positive.

In general, any linear function can be thought of as the sum of a constant term plus a proportional relationship, see Figure 19.

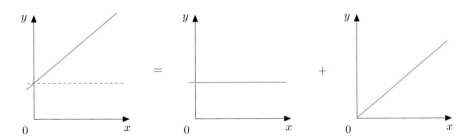

Figure 19

When assuming that a linear function will provide a good enough model, it is often possible to choose the variables carefully so as to obtain a direct proportional relationship. For example, Figure 20(a) shows a spring being stretched by the addition of weights to its end. If you were modelling this situation, perhaps in order to design a weighing device, you could choose the variables as the length of the spring and the weight suspended by it, in which case the graph of the linear function would be as shown in Figure 20(b). However, if you chose the variables as the extension of the spring *beyond* its unstretched length and the weight, you will have a simpler function: a direct proportional relationship. The extension is directly proportional to the suspended weight, as shown in Figure 20(c).

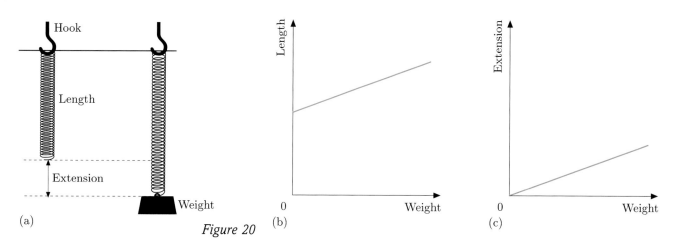

(a)

Figure 20   (b)   (c)

Choosing different variables in this way moves the graph of the function. The mathematical name for this is *translating* the graph; a graph can be translated up or down, or sideways, or both. The graph of $y = mx + c$ is equivalent to the graph of $y = mx$, translated up by c.

Activity 16   *Consolidate your handbook entries on linear functions*

Look back at the notes you made in previous units on linear functions and direct proportional relationships for your handbook. (*Units 2, 6, 7,* and *10* may have relevant notes.) Make sure that you have good notes on linear functions, direct proportionality and the connection between them. Add to them if necessary and try to write in a form which explains to someone else (or yourself, sometime in the future) the meaning of the terms 'linear' and 'directly proportional', as well as including the relevant equations and graphs. If possible, read out your notes to someone else and ask them to explain the concept back to you, to test how well you have communicated your ideas.

You will need your Learning File for other activities in this section.

Activity 17   How did you find this consolidation?

Spend a moment thinking about how you went about the last activity. Did you find it easy to consolidate your notes? Looking back could you understand your previous writing? Did your previous notes fit together well or did you feel you had to rewrite most of them? Do you now feel that you understand linear functions and direct proportionality well and that you have a good concise summary? Are there aspects of the way you went about this activity that will help you to tackle the consolidation of your notes on other functions in subsequent subsections? If so, you could add them to the sheet used in Activity 1.

4.2   Quadratic functions

The simplest quadratic function is of the form:

$$y = ax^2 \qquad (a \text{ a non-zero constant})$$

The basic graph is a parabola, with its vertex at the origin $(0,0)$ with the *sign* of a determining which way up it is and the *size* of a determining how broad or shallow it is. This particular parabola has the y-axis as its axis, about which it is symmetrical.

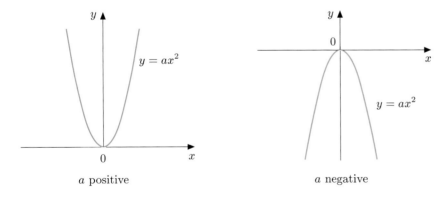

Figure 21

The equation expressing a square proportional relationship is $y = ax^2$, where a is the constant of proportionality (also called the *scale factor*, as it stretches the graph in one direction).

In many square proportional relationships, a and both the variables will be positive and so only part of the parabola is relevant.

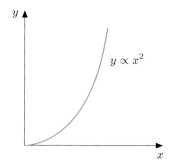

Figure 22   Square proportionality

The standard parabola can be translated up and down and to the left or right. When this happens the associated quadratic function changes accordingly. If the parabola is moved up by l and to the right by k (so its new vertex is at (k, l)), then the equation becomes

$$y - l = a(x - k)^2$$

or $$y = a(x - k)^2 + l$$

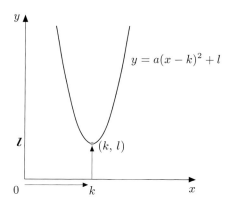

Figure 23

When this is multiplied out and the terms collected together, the equation is of the form:

$$y = ax^2 + bx + c$$

where a, b and c are parameters which specify which particular quadratic function you have. The parameter a determines both which way up and how broad or narrow the parabola is, and c is the y-intercept (the value of y when $x = 0$).

Parabolas are useful in modelling situations where quantities rise to a maximum value or peak and then fall again, or fall to a minimum value and then rise again.

Activity 18   *Reminder of the position and shape of parabolas*

On your calculator, plot the graphs of the following quadratic functions and comment on their shape; for example, the position of their vertex, whether it gives a maximum or minimum value, and how wide or narrow it is in comparison with the standard $y = x^2$ parabola.

(a) $y = x^2$ (b) $y = 0.5x^2$

(c) $y = 2x^2$ (d) $y = 2(x - 1)^2 + 2$

(e) $y = -x^2$ (f) $y = -0.5x^2$

(g) $y = -2x^2$ (h) $y = -2(x - 1)^2 + 2$

Another common use of quadratic functions is in situations when the rate of change of a variable can be modelled as a linear function, so the variable itself can be modelled by a quadratic function. An example is the case where velocity (rate of change of position) is modelled as changing linearly (with a constant gradient, which is the acceleration): position will then be given as a quadratic function of time. You may remember that a quadratic function fitted the braking distances from the *Highway Code* earlier in the *Calculator Book* (based on a model of constant braking), and a quadratic function was also used for the falling car in the video section of *Unit 11* (which was based on a model of constant acceleration due to gravity).

Activity 19   *Consolidate your handbook notes on quadratics, parabolas and square proportion*

Look back at the notes you made on quadratics and parabolas in *Unit 11* and on square proportionality in this unit, together with your answers to Activity 18. Consolidate them into good, clear, concise notes about the relationship between the important features of the graph of a parabola and the associated quadratic function. Also note down the links you see between square proportionality and quadratic functions.

4.3 *Cubics and other positive power functions*

You met cubic proportional relationships earlier in this unit. They were relationships of the form:

$$y \propto x^3$$

which can be expressed in an explicit equation as

$$y = ax^3$$

where a is the constant of proportionality. This function leads to a graph which is shown in Figure 24, part of which you have met before for the cubic proportional relationship (where x and y were both positive).

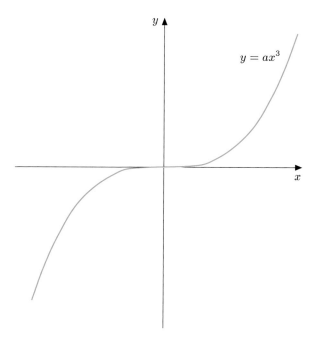

Figure 24   Graph of $y = ax^3$

As in the case of the parabola, the value of a determines how *broad* the *cubic* is, and the sign of a determines which way up the graph is. However, this graph is not symmetrical about the y-axis. The left-hand side is like the right, but it is upside down, a reflection in the x-axis as well as in the y-axis.

Activity 20   *Plotting $y = ax^3$*

Plot the following graphs on your calculator and comment on the effect of the parameter a in $y = ax^3$.

$$y = x^3 \quad y = -x^3 \quad y = 2x^3 \quad y = 0.5x^3$$

Just as quadratics can be translated, so can cubics. To translate a graph up or down by an amount l, replace y by $(y - l)$; to translate a graph by an amount k sideways to the left or right, replace x by $(x - k)$. (If l is positive, the graph translates *up*; if k is positive, the graph translates to the *right*.)

Activity 21   *Translating cubics*

Plot the following cubics and check the above statement about translating graphs.

(a) $y = x^3$

(b) $y = (x - 1)^3$

(c) $y = x^3 + 2$ (which is the same as $y - 2 = x^3$)

(d) $y = (x - 1)^3 + 2$.

It is perhaps interesting to note that this is not the only type of cubic function. Certain cubics are not simply translations of the cubic proportional relationship—something which is true for quadratic functions. The general equation for a cubic is:

$$y = ax^3 + bx^2 + cx + d$$

where a, b, c, d are numerical parameters.

Activity 22   *Maximum and minimum*

On your calculator draw the graphs of the following cubic functions for values of x between -2 and $+2$.

(a) $y = x^3 - x$

(b) $y = x^3 + x$

Comment on the number of maximum and minimum points and the number of times each graph crosses the x-axis.

Activity 23   *Add cubics to your handbook*

Look back at any notes you made on cubic proportional relationships and add to them some notes about the graphs of cubic functions.

4.4 Polynomial functions

Linear functions, quadratic functions and cubic functions are part of a bigger family of functions called *polynomials*. These are functions which involve sums of powers of x (the independent variable): linear functions involve a term in x to the power one; quadratics involve a term in x to the power two (x squared); cubics involve a term in x to the power three.

The Greek prefix 'poly' means 'many'.

Next come *quartics* which also involve a term in x to the power four and *quintics* which involve x to the power five.

Another way of expressing this is as follows: polynomials are functions which are the sum of terms involving (whole number) powers of x (the independent variable). The first polynomials are as follows where $a, b, c, d, \ldots$ represent any numbers and $a \neq 0$.

Linear $\qquad ax^1 + bx^0$

Quadratic $\quad ax^2 + bx^1 + cx^0$

Cubic $\qquad ax^3 + bx^2 + cx^1 + dx^0$

Quartic $\qquad ax^4 + bx^3 + cx^2 + dx^1 + ex^0$

Quintic $\qquad ax^5 + bx^4 + cx^3 + dx^2 + ex^1 + fx^0$

The degree of a polynomial

Polynomials are useful for modelling; they have some interesting properties.

In general, a polynomial of degree n is the sum of terms with powers of x up to x^n (n being a positive number). So linear functions are polynomials of degree 1; parabolas are polynomials of degree 2; cubics are polynomials of degree 3; and so on. Below in the box is a result that is true about all polynomials.

This is an example of a statement of a general mathematical theorem about *all* possible polynomials—the reasoning and the proof are beyond the scope of this course.

> A polynomial of degree n crosses the x-axis up to n times and has at most $(n - 1)$ turning points (maximum and/or minimum points).

Activity 24

Write out the above statement for the following values of n and check that it holds in these cases. Try different values of a, b, c, d.

(a) $n = 1$ (b) $n = 2$ (c) $n = 3$ (d) $n = 0$

Remember that the rate of change of a quadratic function (a polynomial of degree 2) is a linear function (a polynomial of degree 1).

The rate of change of a linear function (a polynomial of degree 1) is a constant (a polynomial of degree 0).

The rate of change of a cubic function (polynomial degree 3) is a quadratic function (polynomial degree 2), and the rate of change of a quartic function (degree 4) is a cubic (degree 3). And so on.

Here is another general statement about *all* polynomials.

> In general, the rate of change of a polynomial (of degree n) is another polynomial of one degree less (degree $n - 1$).

Polynomial functions can be used in modelling quantities which have a number of maximum and minimum points. There is often a choice of function to use in a particular modelling situation and polynomial functions may be used in preference to more complicated functions. They can also be used to approximate complicated functions over small ranges.

Polynomial approximations

If you plot $y = \sin x$ on your calculator, in radian mode, for the range of x from 0 to 1 and on the same graph plot $y = x$, you will see that the two graphs are quite close from $x = 0$ up to about $x = 0.5$. Hence over this small range the function $y = x$ might be used as a very good approximation to $y = \sin x$. The approximating function $y = x$ is a much simpler function than $y = \sin x$, so for some purposes it might be preferable to use the simpler approximation rather than the actual function. If you wanted a better polynomial approximation, you could add other higher powers of x. If you plot $y = x - \dfrac{x^3}{6}$ you would find it hard to distinguish between it and $y = \sin x$ over the range $x = 0$ to $x = 1$. However, if you zoomed out, you would find that for values of x greater than 1, the two graphs get further apart, and beyond $x = 1.5$ they are very different. Adding higher powers can get closer to $\sin x$ for a larger range of x; for example, $y = x - \dfrac{x^3}{6} + \dfrac{x^5}{120}$ is very close to $y = \sin x$ from $x = -2$ to $x = 2$.

Cos x can be approximated by $y = 1 - \dfrac{x^2}{2}$.

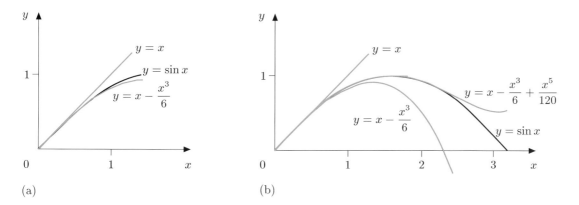

$$(a)$$ $$(b)$$

Figure 25 $y = \sin x$ and some polynomial approximations

Example 8 *Approximating e*

One use of polynomial approximations is to evaluate the functions which would be long or difficult to evaluate from the function itself.

For instance, the function $y = e^x$ can be approximated by the polynomial function.

$$y = 1 + x + \frac{x^2}{2} + \frac{x^3}{6} + \frac{x^4}{24} + \frac{x^5}{120} \tag{11}$$

How closely does this function approximate $e^{0.0453423123}$?

With $x = 0.0453423123$, both $y = e^x$ and the polynomial approximation given in (11) give 1.046385989. So the approximation and the function give the same value correct to ten figures displayed on a calculator.

In fact, you may wonder how your calculator works out values such as $e^{0.0453423123}$. The definition of an exponential function like this means that the function $e^{0.0453423123}$ is defined as 'multiply e by itself $453\,423\,123$ times and then find the $10\,000\,000\,000$th root'. If the calculator tried to do this it would take a very long time and it would probably exceed its memory capabilities.

Hence calculators use other methods of calculating the value of functions like this including using polynomial approximations to the functions instead of the actual function. Although the polynomial approximation for $y = e^x$ given in (11) works well for small values of x, it does not work well for large values of x, and so a calculator may need a number of different polynomial approximations to be able to evaluate a function like $y = e^x$ over the whole range of values of x likely to be entered.

The value of e is actually calculated from approximations like that in (11), as e is just e^1. If you put $x = 1$ in the approximation function (11), you will get an approximation for e of:

$$1 + 1 + \frac{1}{2} + \frac{1}{6} + \frac{1}{24} + \frac{1}{120} = 2.71666667\ldots$$

You need many more terms of a polynomial like this before you get the same degree of accuracy for e that your calculator gives.

There are similar polynomials which can give approximations to the value of π and one test of the power of a computer is to how many figures it can accurately calculate π.

Activity 25   Notes on many nomials

Write a handbook entry on the meaning of the term 'polynomial function' with some examples, their properties and their uses.

4.5 Negative power functions

You have met inverse proportional relationships of various kinds. For instance, if y is inversely proportional to x, then $y \propto x^{-1}$, which means:

$$y = kx^{-1} = \frac{k}{x}$$

where k is the constant of proportionality. You have also met inverse square proportionality relationships written $y \propto x^{-2}$ which means:

$$y = kx^{-2} = \frac{k}{x^2}$$

These inverse proportional relationships are usually only valid for positive x, y and k and so the graph of the proportional relationship is only part of the graph of the whole function. Functions involving other negative powers of x have similarly shaped graphs to those of $y = kx^{-1}$ and $y = kx^{-2}$, as shown in Figure 26 below.

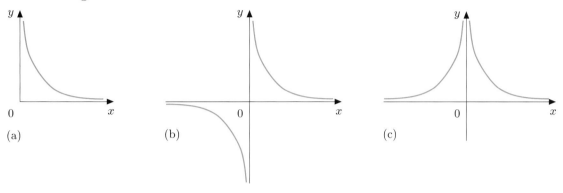

Figure 26   (a) y inversely proportional to a power of x (b) $y =$ negative odd power of x (c) $y =$ negative even power of x

Look at these functions and in particular think about what happens near $x = 0$.

▶   What happens in the simplest case, $y = \dfrac{1}{x}$, as x gets closer to 0?

As x gets closer to 0, $1/x$ gets bigger, but since you cannot divide by 0, there is no value of y actually corresponding to $x = 0$. Although the graph gets very close to the y-axis, it does not actually touch it. In mathematical language, you can say that the function is *not defined* at $x = 0$ and that the y-axis is an asymptote: the value of y gets bigger (either positive or negative) the closer to the y-axis you get, but it has no value for $x = 0$. So the function never crosses the y-axis, it just gets closer and closer to it. It is said to behave *asymptotically*.

Also, as x gets bigger and bigger (either positive or negative), the value of the function gets closer to zero. The function does not cross the x-axis, but gets closer and closer to it asymptotically.

Activity 26   *Inverse relationships*

Write a handbook entry on inverse proportional relationships and their graphs for your handbook.

4.6 Exponential functions

A different type of function from the proportional relationships is the exponential. It is worth distinguishing positive exponentials and negative exponentials, whose functions are given by the following equations and whose graphs are shown in Figure 27.

$$y = a \exp(kx)$$

or

$$y = ae^{kx} \qquad (k \text{ positive})$$

and

$$y = a \exp(-kx)$$

or

$$y = ae^{-kx} \qquad (k \text{ positive})$$

Note that $\exp(x)$ is the same as e^x, and ae^{kx} is the same as ab^x, where $b = e^k$ (because $e^{kx} = (e^k)^x = b^x$).

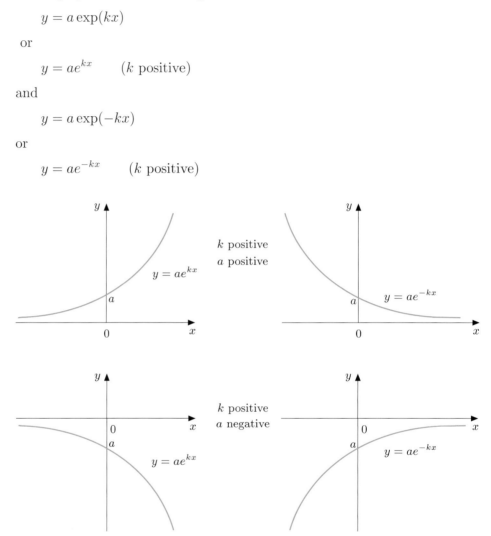

Figure 27   Positive and negative exponentials, a positive or negative

Note that the y-intercept a is where the graph crosses the y-axis; that is, the value of y for $x = 0$. It can be positive or negative.

▶   How do these graphs differ from those of proportional relationships (direct or inverse) or polynomials?

The graphs do not go through the origin, as direct proportional relationships do; but although they do not cross the x-axis, they do cross the y-axis (unlike inverse proportional relationships). They have no turning points, but they do approach asymptotically to either the positive x-axis (negative exponentials) or the negative x-axis (positive exponentials).

The positive exponential function is good at modelling unchecked growth, where there is no limiting factor and the population grows proportionately to its size. However, this is rarely a good model indefinitely; for example, exponential growth is quite a good model for the early stages of a chain letter, but the number of letters in circulation must reach saturation point sometime.

The negative exponential is good for modelling decay—quantities like radioactive substances which gradually get closer to zero.

However, because the negative exponential functions can be translated up or down in similar ways to the quadratics, they also provide good models for growth or decay to an equilibrium level or limit.

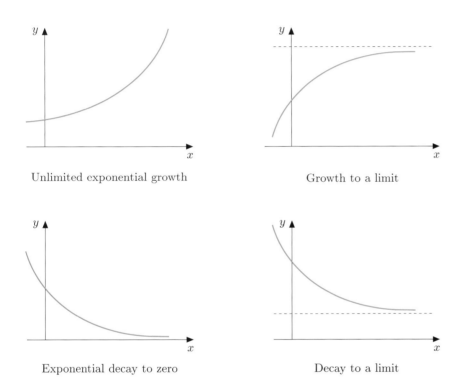

Unlimited exponential growth   Growth to a limit

Exponential decay to zero   Decay to a limit

Figure 28   Growth and decay

Activity 27   *Growth and decay*

Draw the graphs of the following functions on your calculator.

(a) $y = 3 - 2\exp(-0.5x)$

(b) $y = 3 + 2\exp(-0.5x)$

(c) $y = 2\exp(-0.5x)$

(d) $y = 2\exp(0.5x)$

(e) $y = 3 - 2\exp(0.5x)$

(f) $y = 3 + 2\exp(0.5x)$

Which might be good models of:

◇   unlimited growth;

◇   growth to a limit;

◇   decay to nothing;

◇   decay to a limit?

Another function you met in relation to the exponential function was the logarithmic function which 'undoes' the exponential function, and so exponential and logarithmic functions are called *inverse functions*. For example, the function $y = \log_e x$ undoes the function $\exp(x)$.

The graph of the function $y = \log_e x$ is also shown in Figure 29. Note it is only defined for positive values of x and it is asymptotic to the y-axis.

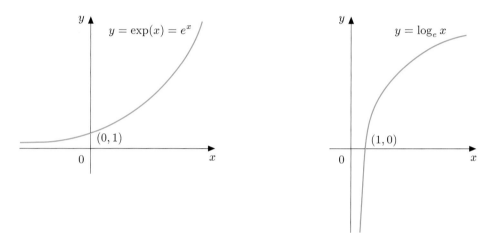

Figure 29

Notice that the graph of the logarithm function is like that of the exponential, but with the axes interchanged, see Figure 30.

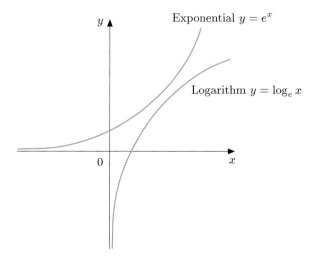

Figure 30

The same is true for other pairs of functions which are inverses. (For example, $y = x^2$ and $y = \sqrt{x}$; $y = \sin x$ and $y = \sin^{-1} x$; $y = x^{-1}$ and $y = x^{-1}$.)

Activity 28   Log and exp

Look at any notes you made previously for your handbook on the exponential and logarithmic functions and add to them if you feel it is necessary.

4.7   Fitting models to data

The above mathematical functions can be used to model given sets of data using the regression facilities on your calculator. Section 13.2 of the *Calculator Book* provides a summary of the different regression facilities and their use. You should study this section now, consolidate and extend any notes you have made previously about regression, in order to have a good clear handbook entry on the subject.

Now study Section 13.2 of Chapter 13 in the Calculator Book.

4.8   Periodic or trigonometric functions

The type of function suitable for modelling vibrations or oscillations, such as those in the air when a pure note is sounded, is trigonometric and the simplest such function is:

$$y = \sin x$$

Others you have met in *Unit 9* are:

$$y = \cos x$$

$$y = \tan x = \frac{\sin x}{\cos x}$$

The function $\sin x$ varies between a maximum of $+1$ and a minimum of -1. The family of functions

$$y = a \sin x$$

is just this function scaled to have maxima and minima of $+a$ and $-a$; a is called the *amplitude*.

The family of functions

$$y = a \sin bx$$

is similar, but the period depends on the value of b. The period here can be thought of as the *time* for one complete cycle.

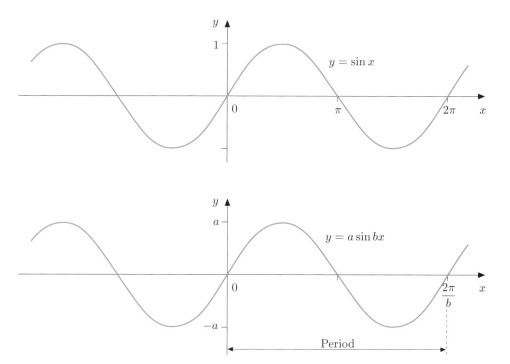

Figure 31

Such functions are used for modelling oscillations.

Recall the investigation from Chapter 9 of the *Calculator Book*.

The period of an oscillation and the frequency of the oscillations both depend on the parameter b: the period is $2\pi/b$ and the frequency is $b/2\pi$.

The sine function can be translated in a similar way to a parabola. Translating sideways models what is called the *phase* of the oscillation. The phase (where in the cycle $y = 0$ occurs) depends on the parameter c in $y = a\sin(bx + c)$.

These phenomena are explored further in *Unit 15*.

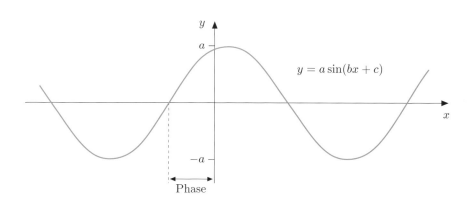

Figure 32

Activity 29   *Phase, frequency and amplitude*

Draw the following graphs on your calculator and state which one has the greatest amplitude, which the greatest frequency and which ones have the same phase as each other.

(a) $y = 4\sin x$

(b) $y = 2\sin 2x$

(c) $y = 2\sin(x + \pi/4)$

(d) $y = 6\sin(0.5x + \pi/4)$

Activity 30   *Trigonometric functions*

Look back at any notes which you have made for your handbook on trigonometric functions and add to them if you feel it would be useful.

4.9   *Functions involving more than one variable*

Some of the proportional relationships you saw earlier in this unit involved more than one variable, like the radius of the cake and its height. In these cases, you took the volume as fixed in order to express one variable in terms of the other. However, you may sometimes want to express a proportional relationship in terms of more than one variable. This is sometimes called *simultaneous proportionality*. An example of this is the volume of a circular-base cake.

The volume is proportional to its radius squared and to its height.

$$V \propto r^2$$
$$V \propto h$$

Put together this can be expressed as a proportional relationship involving more than one variable (two variables actually: r and h).

$$V \propto r^2 h$$

In fact, the constant of proportionality is π, and so

$$V = \pi r^2 h$$

The surface area of the side of such a cake is given by

$$S = 2\pi r h$$

and so in this case S is directly proportional to both r and h.

Your calculator cannot draw graphs of functions of more than one variable. Such graphs are three-dimensional not two-dimensional. However, there are computer packages which will draw such graphs in a similar way to the Ordnance Survey representations of landscape in the maps in *Unit 6*. The work you did on maximum points, minimum points and saddle points there is just as applicable to graphs of functions of two variables as to landscape. The similarity goes further, and functions of two variables can be represented by contour maps in the same way as landscape features.

Activity 31 *Proportional relationships*

Add a note in your handbook about proportional relationships which involve more than one variable, and representations of functions of two variables. Look back at any notes you made in *Unit 6* on contours, maximum and minimum points, and saddle points and add to them if you wish.

4.10 Inequalities

So far in this section you have only been looking at equalities: for instance, $y = mx + c$, or $y = \sin x$. The general notation for such functions is $y = f(x)$, which is said aloud as 'y equals f of x' or 'y equals some function of x'. However, inequalities are also important (in modelling). The range of values a variable can take can be concisely expressed as an inequality. For example, if only positive values of the time t are relevant, then this can be expressed as $t > 0$. Sometimes a situation involves a maximum or a minimum constraint—a function must be less than a specified amount (for example, the amount that can be spent must be less than a specified sum of money) or at least a certain quantity of goods must be ordered.

In these situations, an inequality is appropriate. Graphically, inequalities can be represented as shaded areas on either side of the graph of the equality as shown in Figure 33. The equality is where $y = f(x)$; above the graph $y > f(x)$, and below the graph $y < f(x)$.

Activity 32 *Notes on inequalities*

Look back at any notes you made on inequalities in *Unit 10* and add to them if necessary.

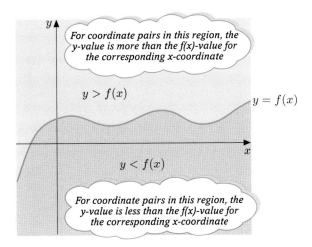

Figure 33   Inequalities and equalities

4.11 *Discrete and continuous models*

The range of values for which a function makes sense may depend on the modelling situation or on mathematical considerations. Sometimes, it only makes sense for a variable to be a whole number: for example, numbers of breeding seasons for birds. In such cases, the variable only takes discrete values. In other cases, it makes sense for the variable to take any value in a continuous range: for example, the distance travelled on a journey.

Sometimes a model may take this into consideration. For instance, when planning a system for loan interest and repayment, the discrete nature of the repayments and charging periods for interest needs to be built into the final model. However, sometimes in early models it is easier to ignore the discrete nature of the variables.

In other cases, the reverse may be true. Radioactive decay, for example, is often discussed in terms of half-lives—the period at the end of which half the radioactive substance will have decayed—as if time went by in discrete half-life jumps. You need to be aware of the range of values which a variable can take in a model or a mathematical function in any particular situation; and whether this is a continuous range of values or a discrete set of values within a range.

In the *Calculator Book*, you were alerted to problems which can arise in interpreting graphs because of the discrete nature of the display, in that the screen is made up of a large number of discrete dots or pixels, which are either 'on' or 'off'. So when displaying a graph on the screen, a discrete set of points represents a continuous function.

Activity 33 *Notes on discrete and continuous quantities*

Look back at any notes you made previously in *Units 5, 7* or *12* about discrete and continuous quantities. Are they sufficient or do you need to add to them?

This entire section has been concerned with mathematical consolidation. You should now have a good repertoire of functions to use in many situations. However, you are not expected to remember everything by heart—you should also now have a good set of reference notes in your handbook to use in the rest of this course and beyond.

Outcomes

After studying this section, you should be able to:

◇ recognize and describe, to someone else, with the help of your handbook notes, aspects of the following mathematical functions: linear, quadratic, polynomial, negative power functions; exponential, logarithmic, trigonometric, functions of more than one variable; and inequalities (Activities 16 to 32);

◇ describe graphical representations of the above and recognize whether a quantity is discrete or continuous (Activity 33).

5 Modelling using your library of functions

Aims   This section aims to show you a number of modelling situations in which different functions are used from your library of mathematical functions. You would not be expected to produce models like this yourself, but you should aim to follow the models, to be able to explain the main points in your own words as if to someone else, and to comment on the use of mathematical models in the given situation.   ◇

In the television programme *Designer Rides*, modelling was used in the design of the fun-fair rides: positions, velocities, accelerations, jerks and jounces were all designed according to models of thrilling rides. The purpose of these models was to aid design.

In the television programme *Deadly Quarrels*, modelling was used to understand the build-up of the arms race and the way in which the size of opposing sides in warfare affected the killing rates of the opposition and hence the likely outcome of the battle. The purpose of these models was understanding and prediction.

The television programme *Blue Haven* showed models aimed at improving understanding of the way in which hunting affects whale populations. They were models for prediction, but they were models to be used in support of arguments; opposing sides in the ban on hunting debate used models differently.

The purposes of modelling are many and this section is designed to give you a glimpse of a variety of purposes. Each subsection deals with a different modelling context.

Some modelling is done with a design purpose, like the design of a comfortable lift (Subsection 5.3) or the shape of a can (Subsection 5.2). Some modelling is done for understanding and predicting, like 'No King Kongs' (Subsection 5.1) and some is for supporting opposing arguments, such as the average speed of *Bluebird* (Subsection 5.4).

As you study these models, also think about other contexts where you have come across modelling. In the future when you encounter mathematical models, try to identify the purpose behind the model and hence what has been stressed and what ignored in modelling for this purpose.

5.1 No King Kongs

Activity 34   A model for King Kong

Study the reader article 'Sorry no King Kongs', identifying the mathematical arguments as you read and making a note of any assumptions implicit in the argument.

Write down a brief explanation of the arguments used in the article. Which assumptions are used and which mathematical models? Add further comments about the article if you wish.

Recall Hope and John Clearwater's discussion from the very first reading in *Unit 1* about cabbages and spheres.

The assumptions about scaling are the same as for scaling up cake recipes, using cubic and square proportional relationships for volume and area respectively, which seem reasonable. The assumption that mountains are conical shapes is a rough one, but from the experience of cakes, the same proportional relationships are likely to hold for scaling up other shapes. In some ways, taking a cuboid mountain might have made the mathematics easier.

The volume of a cone is used (this is a function of two variables, radius and height) and only mountains of the same radius base as Everest are considered. There are flaws in the argument. The argument holds only for mountains on the same size base as Everest, where the weight will be proportional to the height. No argument is used for mountains with bigger bases.

The assumption that the way in which mountains fall is due to the rock below not being able to support them does not seem exactly foolproof, and specifying where the base of a mountain lies exactly is a tricky question. Similar arguments for animals or monsters seem more feasible. Mountains crumbling by rocks falling off at the side is not considered, but this type of argument may be included in 'other physical considerations' near the end which reduced the maximum height from 100 miles to 15 miles! Hence it is not clear what the purpose of the article was, except perhaps to amuse the reader.

The units used are mixed: miles, feet, pounds; and the term 'quadrillion' is used to denote 10^{15}.

5.2 Can sizes

Manufacturers of cylindrical cans have a choice in the ratio of the height to radius of the can for a given volume. However, a manufacturer may not wish to use more material than necessary and so may choose to minimize the surface area of the can.

Activity 35   *Volume of a can*

A manufacturer has to put a pre-determined volume V of drink, or other substance, into a cylindrical can: however, the proportions of the can vary. How is the height h of the can related to its radius r?

The optimum shape of a half-litre can

Suppose that a can is to contain 0.5 litre ($500\,\text{cm}^3$) and that its radius is $r\,$cm and its height is $h\,$cm.

The volume $V\,\text{cm}^3$ and surface area $A\,\text{cm}^2$ are given by:

$$V = \pi r^2 h = 500 \tag{12}$$
$$A = \text{area of top and bottom} + \text{area of side} = 2\pi r^2 + 2\pi r h \tag{13}$$

From equation (12),

$$h = \frac{500}{\pi r^2} \tag{14}$$

Substituting this in equation (13) gives:

$$A = 2\pi r^2 + \frac{1000}{r} \tag{15}$$

Plotting and tracing this function gives a minimum value at $r \simeq 4.3$, and substituting this back into equation (14) gives:

$$h = 8.6$$

So the most economical can has a height of 8.6 cm and a radius of 4.3 cm (or equivalently a diameter of 8.6 cm).

Activity 36   *What has been assumed?*

Re-read the discussion above about the optimal shape of a can. Identify the assumptions that are made in this argument.

Use your calculator to plot and trace the function given in equation (15), and check that there is a minimum where predicted.

Comment on the predictions in the light of the shape of cans with which you are familiar.

What is the purpose of modelling like this?

5.3 Designing comfortable lifts

This is a light piece of reasoning, which might give you something to think about next time you are travelling in a lift or getting impatient waiting for one. See how much of the principle of the argument you can understand.

A comfortable lift journey

▶   When you travel in a lift, what makes for a comfortable ride between adjacent floors?

Obviously, you want to set off and stop at the appointed position, but if the velocity at which you travel between floors were constant, then you might not get a very comfortable ride. This is because of the sudden change in velocity at the beginning and end of your ride. Velocity is the slope of the graph of position against time, so this can be appreciated better graphically. Look at Figure 34, which shows the position–time graph and the velocity–time graph (sometimes called a 'velocity profile' by lift manufacturers) for a constant-velocity model.

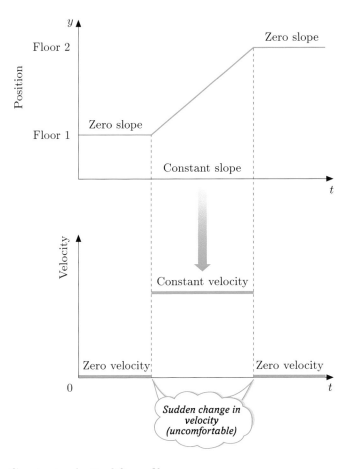

Figure 34   Constant-velocity lift profile

The rate of change of velocity is called *acceleration*, and for a comfortable ride the acceleration must not change suddenly. Acceleration is the slope of the graph of velocity against time. In Figure 34, there are two sudden changes in velocity, making for very uncomfortable starting and stopping.

If there is a sudden change in acceleration you get a jerk. This is used as the technical term for the rate of change of acceleration. Jerk is therefore the slope of the graph of acceleration against time.

The rate of change of jerk is jounce (which is related to bounce), and for a really comfortable ride, this quantity too must not have any sudden changes.

Lift designers can consider a number of possible velocity–acceleration–jerk –jounce profiles for lifts. A suitable profile for a comfortable lift might be something like that shown in Figure 35.

If you watched the TV programme *Designer Rides*, you may remember the terms 'jerk' and 'jounce'.

In designing fun-fair rides there needs to be a certain amount of jerk and jounce to make the ride 'fun', but the reverse is true for designing a comfortable lift!

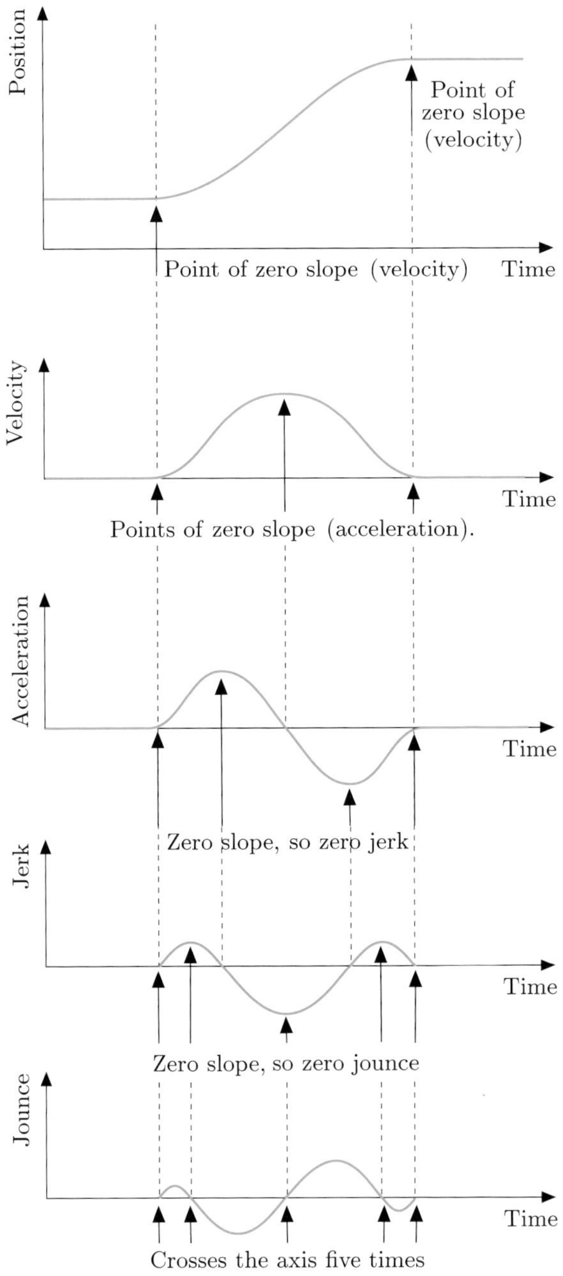

Figure 35

In this profile, the graph for jounce crosses the horizontal axis five times, and so a polynomial of at least degree five would be suitable. Now since jounce is the rate of change of jerk, jerk must be a polynomial of at least one degree more than jounce, that is at least degree six. Similarly, acceleration must be at least one degree more (at least seven) and velocity one degree more still (at least eight). As velocity is the rate of change of position, position must be at least one degree higher; that is at least nine. So in the design of this comfortable lift journey you need to be dealing with polynomials of at least degree nine. It would seem that lift designers must be pretty good mathematicians!

Activity 37 *Lift polynomials*

(a) Look at Figure 34 and explain to someone why this gives an uncomfortable ride.

(b) Look at Figure 35 and explain to someone how the zero points of one graph are obtained from the one above.

(c) Look back at Section 4 to find out why the graph of jounce in Figure 35 must be a polynomial of degree (at least) five.

(d) Check also that the rate of change of a polynomial is another polynomial one degree lower.

(e) Are you convinced of the lift argument or is there another possibility?

There are some notes and graphs from a lift manufacturer in the reader that you may care to look at now. How much can you understand? Can you describe why the velocity profile gives a comfortable ride?

5.4 Average speed of Bluebird

When world speed records are attempted, the speed is calculated according to certain rules. Usually, two runs are attempted over the same course but in opposite directions. An average speed between the runs is then taken as the speed for the purposes of the record attempt. There are a number of ways of modelling the situation and the method used is far from uncontroversial, as the correspondence in the reader article shows.

Part of the reasoning behind the two runs is to counteract any effects of winds or currents. However, there are two equally plausible models for the situation, which are given opposite.

Model 1

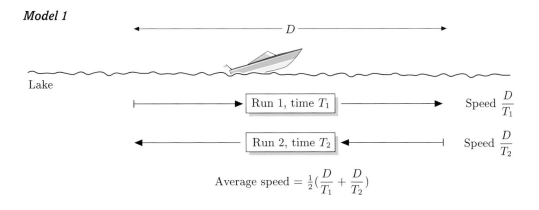

Model 2

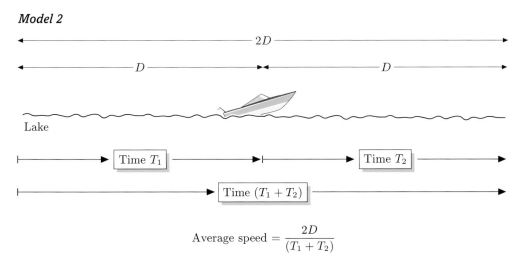

Figure 36

Activity 38   *Modelling to a purpose*

Now read the correspondence about how best to measure the average speed of a boat when it is attempting to break the world speed record over two runs in opposite directions over the same one kilometre course. *Bluebird* is the boat in question. As you read, try to think about the purposes that each letter writer has in mind and to which model they are therefore subscribing.

Explain to someone the different models and hence the different formulas for calculating the average speed of the boat over two runs in opposite directions over the same course.

Note that some of this correspondence refers to the difference between arithmetic, harmonic and geometric means and their properties, all of which you met in *Unit 9*.

Check that if the time on both runs is the same, both formulas give the same result; but that if one time is double the other, then this is not the case.

Which formula do you think would be more favourable for the record breaking attempt?

In summary, models are used for many differing purposes: design; understanding; prediction; economy, supporting an argument; putting over a point (perhaps a teaching point); illustrating or explaining something or sometimes just for fun. In the light of the chosen purpose, some aspects will be stressed, and some ignored, assumptions will be made and results interpreted. You as an onlooker should interpret and evaluate the model in the light of these points. However, it is not always evident from the description of the model what the purpose or assumptions are or what has been stressed and what ignored. Sometimes you need to play the part of a sceptical detective in investigating the ingredients of the model and deciphering the recipe.

Outcomes

After studying this section, you should be able to:

◇   follow, explain and comment on the use of models by others, involving functions, of the type covered in Section 4 (Activities 34 to 38);

◇   read critically accounts of modelling activities, including identifying the purpose, assumptions, and aspects which are stressed and ignored by the model (Activities 34 to 38).

Unit summary and outcomes

This unit has consolidated and also extended a number of mathematical ideas from earlier in the course; these include the concept of proportion and different types of proportion, and the different ways to represent a proportional relationship with symbols, equations, functions and graphs.

Proportional relationships can be represented by power laws and data can be used to fit a power law in particular situations. Your calculator has regression facilities that will do this. The *Calculator Book* summarizes all the regression facilities on the calculator and shows how some of them are related.

The simplest power laws are linear, quadratic and cubic functions. These are also part of a family of functions called *polynomials*, which are often used in approximating other functions for ease of calculation. Other familiar functions in your library, and hence in your Handbook, are the trigonometric family of sine functions, and also exponentials and their inverses, logarithms.

Much of mathematics is concerned with equalities and equations, but in modelling *inequalities* are often more appropriate. Equalities may be represented by lines or curves, and inequalities by areas or regions of the plane. Much mathematics also assumes continuous functions and graphs, whereas some modelling situations involve quantities which take only discrete values.

Modelling is done for a variety of purposes which influence what is stressed and what is ignored, the assumptions made and the interpretations of results. However, much of this is often not made obvious in writing about the modelling. You need to read with a critical detective mind when trying to piece together the whole picture, leading to a clearer understanding.

Activity 39 *Reviewing progress*

Before moving to the next unit, try to allocate some time to review your work before embarking on the last block of the course.

In Block C, you have been encouraged to think about the way you are learning and consciously monitor and review how you are doing. What techniques have you found to be helpful? Do you think the process of reviewing is a valuable part of the learning cycle? Does talking with other students and tutors help you to review and think about your learning? Are there any difficulties that you have encountered (such as time)? A number of activities in this unit have asked you to look back at earlier work and in some cases add further ideas. Does this type of activity help to 'consolidate' your learning? To complete your Learning File activity, include a commentary noting your ideas and thoughts on these or other aspects of your progress. Try to identify what helps you to learn mathematics and what you still want to work on to improve.

Outcomes

After completing this unit, you should be able to:

◇ use proportional relationships in relevant contexts; for example, to scale recipes for different numbers of servings or different-sized containers;

◇ explain the meaning of, and use correctly, words describing proportional relationships; for example, direct or linear proportion, square or cubic proportion, inverse proportion, inverse square proportion, asymptote;

◇ sketch the graphs of and write down proportional relationships symbolically both in the form of a power law equation and as an expression involving the proportion symbol $\propto$;

◇ use your calculator to obtain the best fit power law and other regression functions to a given set of data and interpret the results in practical contexts;

◇ use power regression to test proportional relationships;

◇ manipulate proportional relationships and power-law regression in practical contexts;

◇ recognize and describe, with the help of your Handbook notes, the following mathematical functions: linear, quadratic, polynomial, negative power functions, exponential, logarithmic, trigonometric;

◇ describe graphical representations of the above, explain and comment on their use in models;

◇ read critically accounts of modelling activities, including identifying the purpose, assumptions and aspects which are stressed and ignored by the model.

Comments on Activities

Activity 1

It is sometimes difficult to appreciate the value of spending time on reviewing your work, in order to make good use of the skills that enable effective learning, you need to be aware of the different skills, and when to make best use of them.

As you work through the unit and review your work, it may be useful to make a note of how you are getting on. At the end of the unit, you will then be able to look at your notes to review more objectively. On the other hand, if you are short of time, and this type of activity is not completed, does it matter? In your view, is it a necessary part of the learning process?

Activity 2

For milk (in litres), the constant of proportionality is $\frac{1}{6}$. So

$$a = \tfrac{1}{6}p$$

with vanilla essence (in teaspoons), the constant of proportionality is $\frac{1}{3}$. So

$$a = \tfrac{1}{3}p$$

For eggs, the constant is $\frac{2}{3}$ and so

$$a = \tfrac{2}{3}p$$

The graphs are given in Figure 37

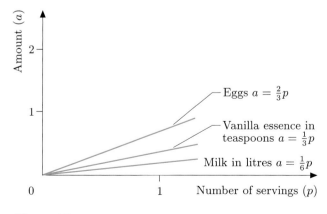

Figure 37

Activity 3

All amounts are given precisely in the table below, but in practice you would round them to nearest convenient amount (1.05 lb would be rounded to 1 lb).

Ingredient	Value of k	Amount for 5 lb cake	Amount for 7 lb cake
Mixed fruit	$\frac{7}{20} = 0.35$	1.75 lb	2.45 lb
Glacé cherries	$\frac{1}{20} = 0.05$	0.25 lb	0.35 lb
Flaked almonds	$\frac{1}{20} = 0.05$	0.25 lb	0.35 lb
Mixed peel	$\frac{1}{20} = 0.05$	0.25 lb	0.35 lb
Plain flour	$\frac{3}{20} = 0.15$	0.75 lb	1.05 lb
Mixed spice	0.25 tsp	1.25 tsp	1.75 tsp
Butter	$\frac{2\frac{1}{2}}{20} = 0.125$	0.625 lb	0.875 lb
Brown sugar	$\frac{2\frac{1}{2}}{20} = 0.125$	0.625 lb	0.875 lb
Eggs (beaten)	$\frac{18}{20} = 0.9$	4.5	6.3
Brandy	$\frac{6}{20} = 0.3$	1.5 tbsp	2.1 tbsp

The size of cake tin and the cooking time would also need adjusting, but not in direct proportion to the weight of the cake.

Activity 4

Instead of r you have $\frac{7}{12}r$. So the volume is $\frac{2}{3}\pi(\frac{7}{12}r)^3$ or $\frac{2}{3}\pi(\frac{7}{12})^3r^3$. So the volume is scaled by a factor of $(\frac{7}{12})^3$ which is 0.1984953704, or about 0.2.

Activity 5

The volume of a cuboid cake of side x is x^3.

(a) The volume of a cuboid cake of side $3x$ is $(3x)^3$, which is $(3)^3x^3$ or $27x^3$. So the volume is scaled by a factor of 27.

(b) The volume of a cuboid cake of side $\frac{x}{3}$ is $(\frac{x}{3})^3$ or $(\frac{1}{3})^3x^3$. So the volume is scaled by a factor of 1/27.

Activity 6

If the area of the top of the cake is A, and its radius is r, then the proportional relationship is:

$$A \propto r^2$$

In the form of an equation, this is:

$$A = kr^2 \quad (k \text{ is constant})$$

Dividing through by k gives:

$$r^2 = \frac{1}{k} A$$

Taking the square root gives:

$$r = \left(\frac{1}{k}\right)^{1/2} A^{1/2}$$

Now $\left(\frac{1}{k}\right)^{1/2}$ is a constant, and so the radius is proportional to the square root of the area of the top of the cake.

Activity 7

(a) Doubling the mass W will mean that the side x is multiplied by $2^{1/3}$ or 1.25992105 or just over one and a quarter. If the original dish size was 120 mm, then 1.26×120 mm is 151.2 mm and so a 150 mm dish would be suitable.

(b) The top surface area will be scaled by $2^{1/3}$ squared which is 1.587401052 or just over one and a half.

(c) The cheese coating would go up from 100 g to 158.7 g which could be rounded to 160 g. So it would not need to be doubled.

Activity 8

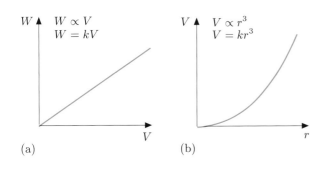

(a)

(b)

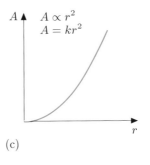

(c)

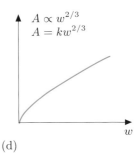

(d)

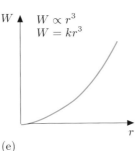

(e)

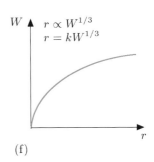

(f)

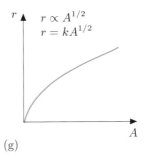

(g)

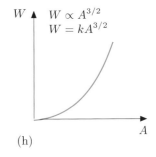

(h)

Activity 9

Your summary will be personal to you, but make sure you have included the three ways of representing a proportional relationship: $y \propto x^n$, $y = kx^n$ and the graph. Also make sure that you have commented on the difference between the graphs for $n > 1$, $n = 1$, $0 < n < 1$, and $n < 0$. You might also have included some examples of where these types of relationship arise.

Activity 10

If eggs were scaled versions of each other, then you would expect length to be directly proportional to diameter and mass to be proportional to length or diameter cubed. The following power regression parameters were obtained from the data.

Variables x, y	Proportionality constant	Power	Regression coefficient
Length, diameter	3.0	0.65	0.96
Diameter, length	0.30	1.4	0.96
Diameter, mass	0.0054	2.5	0.98
Length, mass	0.054	1.7	1.0

It is not clear how much variation there is between eggs, but there is not much support for the theory that eggs of different sizes are scaled versions of each other.

Diameter seems to be proportional to length to the power 0.65 not 1 (or length seems to be proportional to diameter to the power 1.4 not 1); and mass seems to be proportional to diameter to the power 2.5 or length to the power 1.7 rather than 3.

Thus the cubic proportional relationship assumed in Example 2 seems a bit suspect.

The useful data which you now have though is the average mass of eggs of different sizes. This would have been rather difficult to collect in the shop, before deciding which size of egg to buy.

$3\frac{1}{2}$ size 3 eggs will have a mass of about $3\frac{1}{2} \times 65 = 227.5$ grams.

Example 2 suggested that three size 2 eggs or four size 4 eggs were equivalent to $3\frac{1}{2}$ size 3 eggs.

The mass of three size 2 eggs is about $3 \times 70 = 210$ grams.
The mass of four size 4 eggs is about $4 \times 60 = 240$ grams.

Another possibility is three size 1 eggs which is $3 \times 75 = 225$ grams.

So both the suggestions in Example 2 for equivalents to $3\frac{1}{2}$ size 3 eggs are reasonably close, but not as close as three size 1 eggs.

Activity 11

Sarah used a direct proportional relationship between the baking time T and the mass of the cake W, which would give a straight line graph; that is, $T \propto W$.

Judy suggested that the heat H needed to cook the cake would indeed be proportional to its mass W, so $H \propto W$, but that the rate at which the heat entered the cake (H/T) was proportional to the surface area A and so

$$\frac{H}{T} \propto A$$

Since $W \propto H$, this means

$$\frac{W}{T} \propto A \qquad (16)$$

To get T on its own, multiply (16) by T and divide by A, to give:

$$T \propto \frac{W}{A} \qquad (17)$$

However, you know (from Example 6) that

$$A \propto W^{2/3} \qquad (18)$$

and substituting this in equation (17) gives

$$T \propto \frac{W}{W^{2/3}}$$

or

$$T \propto W^{1/3}$$

Allan used his experience and suggested 45 minutes per pound up to three pounds then 8 minutes extra per pound thereafter. This is in effect two linear models: one up to three pounds and another one for cakes bigger than three pounds. The graphs of the three models are shown below.

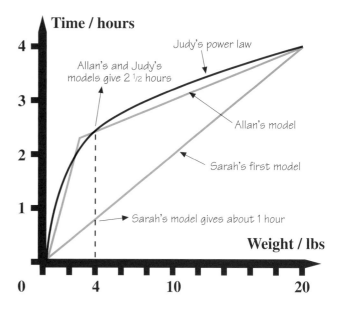

Figure 38 Sarah's, Judy's and Allan's models

Sarah's original model suggested a baking time of about an hour, whereas Allan's and Judy's models suggested nearer two and a half hours.

Activity 12

(a)

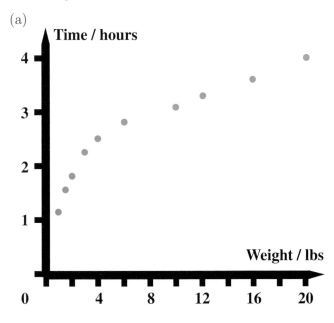

(b) $y = ax^b$ where $a = 1.3$ and $b = 0.38$ (correct to two significant figures)

(c)

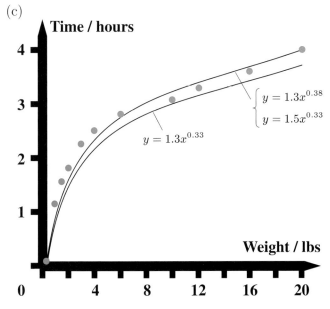

Activity 13

There is little to choose between the best fit regression model and the theoretical power law:

$$y = 1.3x^{0.38} \quad y = 1.5x^{1/3} \text{ (respectively)}$$

and hence either could be used. The choice would depend on the purpose.

Allan's purpose would be to use the model to scale up cooking times for recipes in a cookery book. This would involve a model for the cooking times for scaled up versions of the same shape and type of cake varying with the dimensions of the tin. Judy used the theoretical model to explore the proportional relationships for a circular-based cake.

Activity 14

$$T \propto W^{1/3} \quad \text{and} \quad W \propto r^3$$

so $W^{1/3} \propto r$.

Hence:

$$T \propto r$$

Cooking time is directly proportional to the radius.

Had the regression model been used the relationships would have been as follows:

$$T \propto W^{0.38}$$
$$W \propto r^3$$

So $W^{0.38} \propto r^{0.38 \times 3}$ or $W^{0.38} \propto r^{1.1}$.

Hence $T \propto r^{1.1}$, which is not so simple. Thus, the theoretical model is the simplest for this purpose.

Activity 15

(a) Since $r \propto d$,

$$T \propto d$$

So cooking time is proportional to diameter.

(b) The argument is the same as for a round-based cake with r replaced by x, and so the cooking time is directly proportional to the length of the side of the tin.

(c) All descriptions will be different. However, you should have stated that cooking time is proportional to the dimensions of the tin. So if you double all dimensions of the tin: diameter (or length of the side a square tin) and height, you double the cooking time.

You should also have mentioned that the quantity of ingredients will need to be scaled by this factor cubed. So when you double the tin's dimensions, you scale the ingredients for the cake by 2^3 or 8. You may also have added that any coating (such as icing) needs to be scaled by the square rather than the cube. So if you double the cake tin's dimensions, you scale the ingredients for the coating by 2^2 or 4. A table might help your explanation. Try out your explanation on someone else, perhaps a friend or family member.

(d) When you have written your paragraph, step back from it to look at what it is saying. Is the information given clear and accurate and relevant to the purpose? How have you presented the information? Is it appropriate for your audience? How have you dealt with any mathematical language you have used? Should you include any specialist mathematical terms in a recipe book?

Activity 16

Your entries will obviously be personal to you, but here are some extracts from the notes made by students who were developmentally testing the MU120 teaching materials.

> Linear functions give straight-line graphs $y = mx + c$, where m is the gradient = vertical rise/horizontal run and c the starting value, which is the value of y when $x = 0$.

> Straight lines through the origin represent direct proportional relationships; for example, distance travelled is directly proportional to time taken, when travelling at constant speed.

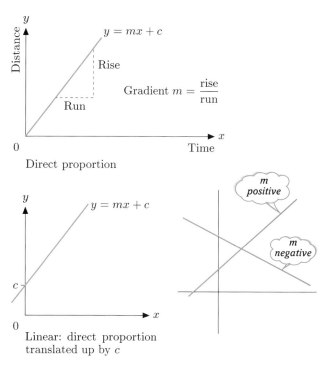

Figure 39

A linear function is the sum of a constant term and a directly proportional term.

Activity 17

Here are extracts from the reflections of some developmental testing students to consolidating and adding to their handbook notes.

> I found this quite difficult: my maths knowledge and experience have never required me to do this before. I had a couple of attempts before someone else could understand my notes. I think I made them too detailed, rather than picking out the main points. However, I enjoyed doing it and found it very valuable in helping my understanding. I now feel more confident about compiling my handbook.

> I found it very difficult to understand my notes and explaining them to somebody else.

> I will use more diagrams in future—they certainly help.

> I realized that I was a bit mixed up about rate of change and the gradient. I had to look back at *Unit 10* to get it clear.

I have never managed to get this type of information so compact before. I'm getting better at note writing. They are now briefer but still have as much information in them.

My mum is getting quite used to me trying out my explanations on her—she is very patient and her listening and asking questions is very helpful to me in clarifying my ideas.

I actually enjoyed this activity. I find that I take in the information much better if I have physically to write it down in my own words. I now feel confident that I understand linear functions quite well.

Activity 18

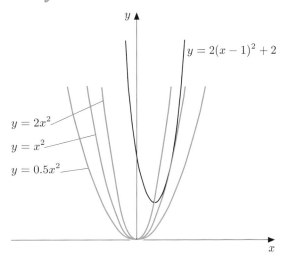

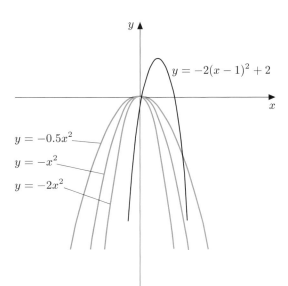

Figure 40

All the parabolas in (a) to (d) open upwards and so have a minimum point; all those in (e) to (h) open downwards and so have a maximum point. (b) and (f) are broader than the standard parabola $y = x^2$ and (c), (d), (g) and (h) are narrower than the standard parabola. However, (a) to (c) and (e) to (g) are all in the standard position with the vertex at the origin. (d) and (h) are translations (2 upwards and 1 sideways) of (c) and (g), respectively with the vertex at $(1, 2)$. Incidentally, (a) to (c) all represent square proportional relationships.

Activity 19

Check that you have included some mention of square proportional relationships being of the form:

$$y = ax^2$$

with a graph which is a parabola (or part of a parabola, if x only takes positive values).

You should have also notes about the graph of $y = ax^2$ being a parabola with the vertex at the origin and the constant of proportionality a. For determining the shape a positive: opening up and giving a minimum point; a negative: opening downwards and having a maximum point; the smaller the value of a the broader the parabola.

You should also have mentioned something about translations of the standard parabola: an example or the general case of $y = ax^2$: translated upwards (or downwards) by l and sideways by k gives the quadratic:

$$y = a(x - k)^2 + l$$

which has its vertex at (k, l). The multiplied-out form of this equation gives the general quadratic function:

$$y = ax^2 + bx + c$$

You should also have written something about when a quadratic model might be appropriate, (for example, when there is a maximum or minimum value) and that the rate of change of a quadratic model is linear. You might have included some examples of motion such as constant acceleration models which lead to linear velocity–time graphs and parabolic (quadratic) position–time graphs.

Activity 20

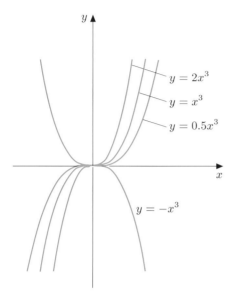

Figure 41

A negative value of a gives an 'upside down' version of the $y = x^3$ graph. For a positive, the larger a is, the narrower and steeper the cubic graph, as in the case of the parabola.

Activity 21

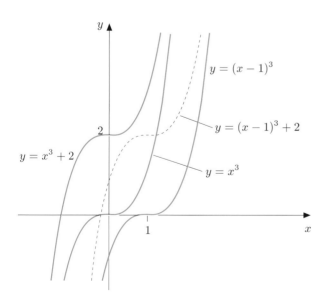

Figure 42

The graphs are translated from (a) the standard graph of $y = x^3$ as predicted:
(b) 1 sideways; (c) 2 up; (d) 1 sideways and 2 up.

Activity 22

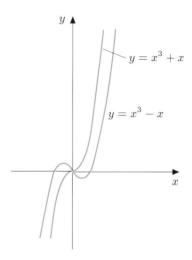

Figure 43

This completes the family of cubic graphs. There may be one minimum and one maximum, that is, two turning points; or there may be no turning points.

Cubic functions will cross the x-axis either once or three times.

Activity 23

You should now be getting into the way of writing clear concise notes for your handbook. Look back over the subsection and check that you have included all the main points about cubic proportional relationships and other cubic functions, including the fact that the graph of a cubic may have two turning points (a maximum point and a minimum point) or no turning points and that the graph will cross the x-axis once or three times. You may also have included the fact that cubics, like parabolas can be translations of the standard cubic which goes through the origin. However, unlike parabolas, they are not symmetric about their axis—you might describe them as anti-symmetric!

Activity 24

(a) A polynomial of degree 1 crosses the x-axis up to 1 time and has at most 0 turning points.

This is a straight line or linear model, which indeed has no turning points and crosses the x-axis once only.

(b) A polynomial of degree 2 crosses the x-axis up to 2 times and has at most 1 turning point.

This is a quadratic function (parabola) which has exactly one turning point and crosses the x-axis at most twice.

(c) A polynomial of degree 3 crosses the x-axis up to 3 times and has at most 2 turning points.

This is a cubic function which actually has at most two turning points and does indeed cross the x-axis at most three times.

(d) A polynomial of degree 0 crosses the x-axis up to 0 times and has at most -1 turning point.

This would be a constant function $y = ax^0$ or $y = a$, which indeed crosses the x axis 0 times but -1 turning points does not make sense and so the generalization is not valid for $n = 0$. Hence polynomials of degree 0, that is constant functions—are often not regarded as polynomials.

Activity 25

You should include the definition of a polynomial as the sum of powers of x, giving examples; you should say what the degree of a polynomial means and give the maximum number of times a polynomial of degree n crosses the x-axis and maximum number of turning points. You might also say that the rate of change of a polynomial function is a polynomial of one degree less. You might also like to include some examples and sketch graphs.

You should indicate that polynomials can be used not only as models in their own right, but also as approximations to other functions perhaps giving the example that functions, such as $y = e^x$, are likely to be evaluated by a calculator using polynomials which approximate the function over particular ranges, because evaluating the function directly is not possible.

Activity 26

You should have mentioned the shape of the graphs (and drawn sketches); that as x gets bigger, y gets closer to 0; and that the graphs do not cross either axis, but are asymptotic to them (and say what asymptotic means).

Activity 27

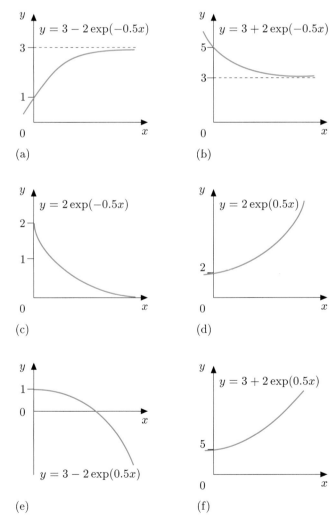

(a) $y = 3 - 2\exp(-0.5x)$

(b) $y = 3 + 2\exp(-0.5x)$

(c) $y = 2\exp(-0.5x)$

(d) $y = 2\exp(0.5x)$

(e) $y = 3 - 2\exp(0.5x)$

(f) $y = 3 + 2\exp(0.5x)$

Figure 44

(d) and (f) model unlimited growth.
(a) models growth to a limit.
(c) models decay to nothing.
(b) models decay to a limit.
Note (e) models none of these situations well.

Activity 28

You should have mentioned the growth/decay aspect of these functions and their use in modelling unlimited growth or decay. Some sketches would be useful, noting that exponential functions do not pass through the origin. You should mention that translations of these functions are useful for modelling growth or decay to a limit—again sketches would be useful to illustrate this. You should also mention that log and exp are inverse functions to one another (and say what this means). You might like to give examples of situations you have met which have been modelled using exponentials.

Activity 29

(d) has the greatest amplitude (6).
(b) has the greatest frequency.
(a) and (b) have the same phase as each other (they they start at the same place on the x-axis).

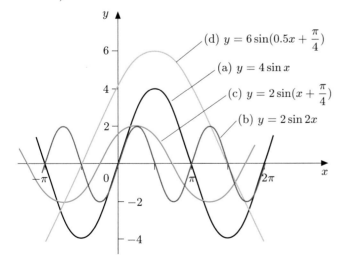

Figure 45

Additional comment
Trigonometric functions, like other functions can be translated up or down, by adding a constant.

The graph of

$$y = \sin x + 2$$

is therefore given by Figure 46.

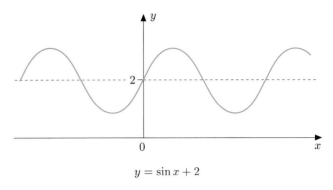

$$y = \sin x + 2$$

Figure 46

The phase can be thought of as a translation sideways.

Activity 30

You should have mentioned the oscillatory nature of such functions and how the parameters (a, b, c) in the general function $y = a\sin(bx + c)$ determine the amplitude, frequency and phase. You might also make mention of translating the graph up or down.

Activity 31

You may like to look back over Sections 1 and 3 and see which other proportional relationships you can find involving more than one variable. You should also make a note about the similarity between representations of landscape and functions of two variables by contours and other representations, giving examples like maximum and minimum and saddle points.

Activity 32

You may have included an explanation of the symbols and a sketch like Figure 33, perhaps using a non-linear function as well as a linear one. Did your notes make sense to you on re-reading them? What have you learned from the experience of taking notes in the earlier subsections to improve your note taking in the future?

Activity 33

You may like to give a number of examples of discrete variables and continuous variables from throughout the course.

Activity 34

The main argument is about scaling up things, like monsters or mountains. Volume and weight vary with the cube of the scale factor, but the cross-sectional area supporting the weight varies with the square of the scale factor. Therefore, there is a limit to the amount of scaling possible. For animals, the strength of bone is the crucial factor, for mountains it is the density (mass per unit volume) and yield strength of the material of which it is made (granite is chosen). A mountain is modelled as a cone on a circular base, and Mount Everest is used as an example to show that this is plausible. The argument is used to predict that the highest possible mountain on Earth (with the same base as Everest) must be less than 100 miles high.

Activity 35

If the volume is V, the height h and the radius r, then for a cylinder

$$V = \pi r^2 h$$

which can be rearranged to give h as the subject:

$$h = \frac{V}{\pi r^2}.$$

As V is constant, this means that h is inversely proportional to the square of the radius r.

Activity 36

The argument assumes that the top and bottom of cans are made of the same material as the sides and no other material is used—for example, to seal the can. It assumes that the quantity of material is the only consideration. It also uses the formulas for the volume and surface area of a cylinder.

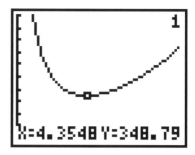

Figure 47   Screen dump of graph of $A = 2\pi r^2 + 1000/r$

The minimum is at about 4.3, but the function is rather flat at the bottom, in particular between 4 and 5.

The ratio of height to radius of cans varies a lot. Drinks cans often have thicker, more robust tops, but have smaller tops (maybe the top is more expensive than the sides). Food cans of this size often have roughly the same height as diameter, but it can vary considerably, so other factors are likely to play a part.

Modelling like this is for economic purposes—manufacturers of cans are in business to make a profit. The above model was quite a simplified one, ignoring such things as strength and thickness of the can, which would need to be taken into consideration in any more refined model. However, it ignores completely the attractiveness of the shape of a can, or the ease of use. (Can somebody get their hand round it to have a drink?) Somebody with a different purpose might have ignored the cost of materials and stressed other features.

Activity 37

(a) The ride is uncomfortable because of the sudden change in velocity at the beginning and end.

(b) After the position graph, every graph is a graph of the rate of change or the slope of the graph above. So a turning point, where the slope is zero, leads to a zero on the next graph, which means the function crosses the axis.

(c) A polynomial of degree n crosses the axis at most n times: the graph of jounce in Figure 35 crosses the axis at least five times so must be given by a polynomial of degree at least five.

(d) The rate of change or slope of a straight line (degree 1) is a constant (degree 0); the slope of a parabola (degree 2) is a linear function (degree 1), and so on.

(e) The argument for jounce suggests that a polynomial is suitable, but it is not the only possibility. A trigonometric function of some type might be another possibility. Different functions might be used to model different parts of the ride; in particular, there might be a period of constant velocity in the middle of the ride. However, it would still need to be a complicated mathematical model and so the statement that lift designers must be pretty good mathematicians is pretty convincing.

Activity 38

The average speed is calculated from two 1 km runs (in opposite directions). One way of calculating the average is to calculate the individual average speeds of each run and then take the mean of these two. (This corresponds to model 1.)

The other way is to divide the total distance (2 km) by the total time to complete both runs. (This corresponds to model 2.)

The letters provide some examples of these methods, but none provide formulas.

However, if the times for the two runs are T_1 and T_2 seconds, respectively, then the first method gives individual average speeds in km per second of $1/T_1$ and $1/T_2$ and the mean of these is

$$A = 0.5(1/T_1 + 1/T_2)$$

or $\quad A = 0.5(T_1 + T_2)/(T_1 T_2).$

The second method gives an average speed in km per second of

$$A = 2/(T_1 + T_2). \tag{19}$$

So the formulas are very different.

However, if $T_1 = T_2$, then the first formula gives:

$$0.5(T_1 + T_1)/(T_1 T_1) = 0.5(2T_1)/(T_1 T_1)$$
$$= T_1/(T_1 T_1) = 1/T_1$$

The second formula gives:

$$2/(T_1 + T_1) = 2/(2T_1) = 1/T_1$$

So both formulas give the same average speed in this case.

If $T_1 = 2T_2$, then the first formula gives

$$0.5(2T_2 + T_2)/2(T_2 T_2) = 0.5(3T_2)/2(T_2 T_2)$$
$$= \tfrac{3}{4}(1/T_2)$$

and the second formula gives

$$2/(2T_2 + T_2) = 2/(3T_2) = \tfrac{2}{3}(1/T_2).$$

The first formula gives a higher result and so is more favourable to the record-breaking attempt.

Activity 39

There are no comments on this activity.

Acknowledgements

Cover

John Regis, sprinter: Press Association; crowd scene: Camera Press; car prices: Edinburgh Mathematical Teaching Group; other photographs: Mike Levers, Photographic Department, The Open University.

Index